The Matter of Death

Also by Jenny Hockey

DEATH, GENDER AND ETHNICITY (*co-edited*)

DEATH, MEMORY AND MATERIAL CULTURE (*co-authored*)

BEYOND THE BODY: Death and Social Identity (*co-authored*)

EXPERIENCES OF DEATH: An Anthropological Account

GRIEF, MOURNING AND DEATH RITUAL (*co-edited*)

Also by Carol Komaromy

DEATH AND DYING: A Reader (*co-edited*)

MAKING SENSE OF DEATH, DYING AND BEREAVEMENT: An Anthology (*co-edited*)

DILEMMAS IN UK HEALTH CARE (*edited*)

COMMUNICATION, RELATIONSHIPS AND CARE (*co-authored*)

Also by Kate Woodthorpe

THE CULTURE AND ECONOMIES OF DEATH

The Matter of Death

Space, Place and Materiality

Edited by

Jenny Hockey
University of Sheffield, UK

Carol Komaromy
The Open University, UK

and

Kate Woodthorpe
University of Bath, UK

palgrave
macmillan

First published 2010 by
PALGRAVE MACMILLAN

Palgrave Macmillan in the UK is an imprint of Macmillan Publishers Limited, registered in England, company number 785998, of Houndmills, Basingstoke, Hampshire RG21 6XS.

Palgrave Macmillan in the US is a division of St Martin's Press LLC, 175 Fifth Avenue, New York, NY 10010.

Palgrave Macmillan is the global academic imprint of the above companies and has companies and representatives throughout the world.

Palgrave® and Macmillan® are registered trademarks in the United States, the United Kingdom, Europe and other countries.

ISBN-13: 978–0–230–22416–2 hardback

This book is printed on paper suitable for recycling and made from fully managed and sustained forest sources. Logging, pulping and manufacturing processes are expected to conform to the environmental regulations of the country of origin.

A catalogue record for this book is available from the British Library.

A catalog record for this book is available from the Library of Congress.

10 9 8 7 6 5 4 3 2 1
19 18 17 16 15 14 13 12 11 10

Printed and bound in Great Britain by
CPI Antony Rowe, Chippenham and Eastbourne

Many dying and bereaved people and their carers participated in the various studies here, in the hope of making a difference. We dedicate this book to those people.

Contents

List of Figures and Table

Figures

Table

Biographical Information

Jan Bleyen is an assistant at the research unit Modernity and Society 1800–2000 at the Katholieke Universiteit Leuven in Belgium. A historian, his main interests lie in anthropological history – emotion, the body, spirituality – in particular, issues dealing with loss. He primarily works on the changing culture of mourning and teaches oral history skills and analysis. In 2005 he published De dood in Vlaanderen. Opvattingen en praktijken in Vlaanderen na 1950 (*Death in Flanders. Beliefs and Practices since 1950*) with Davidsfonds. He has also published several articles on the history of death, including three chapters in the *Encyclopaedia of Death and the Human Experience* (Sage, 2009). His current PhD-project Stillborn: A history of meaning-making focuses on performativity.

Andy Clayden is Senior Lecturer at the Department of Landscape, University of Sheffield. He is also a practising landscape architect who specialises in aspects of sustainable design. His research interests focus on the design and management of cemeteries and specifically natural burial. He has co-authored books on different aspects of sustainable landscape design and has also contributed to the development of government guidance on natural burial and published refereed articles and book chapters on this subject.

Douglas J. Davies trained in both anthropology and theology and is currently Professor in the Study of Religion at Durham University and Director of The Durham Centre for Death and Life Studies. His books include, *The Theology of Death* (2008), *The Encyclopedia of Cremation* (2005 with Lewis H. Mates), *A Brief History of Death* (2004), *An Introduction to Mormonism* (2003), *Death, Ritual and Belief* (2002), *Reusing Old Graves* (1995 with A. Shaw) and *The Mormon Culture of Salvation* (2000). He is an Oxford D.Litt., holds an Honorary Dr Theol. from Sweden's Uppsala University, and is an academician of the Academy of Social Sciences.

Clare Gittings is the author of *Death, Burial and the Individual in Early Modern England* (Croom Helm 1984), and co-editor of *Death in England: An Illustrated History* (Manchester University Press 1999). Several of her published articles concern visual representations of death and dying; she works at the National Portrait Gallery.

Sheila Harper completed her PhD at the University of Bath, UK, in 2008, and is now a postdoctoral researcher at the University of Sydney, Australia. Her research interests focus on the public perception of dead bodies in different contexts.

Jenny Hockey trained as a social anthropologist and is Emeritus Professor of Sociology at the University of Sheffield. She has published widely on ageing, death, dying, disposal, ritual, bereavement and memorialisation. She is president of the Association for the Study of Death and Society (ASDS) and her most recent research has explored the retention of ashes after cremation for independent disposal and the cultural, social and emotional implications of natural burial.

Leonie Kellaher is Emeritus Professor at The Cities Institute, London Metropolitan University. As a social anthropologist her work has centred on the qualitative study of environments; firstly, special settings for older people and more recently, environments of death, notably cemeteries. The material culture of the built, landscaped and natural elements making up both these settings is a particular focus.

Carol Komaromy is a medical sociologist who has worked as an academic at the Open University since 1994 and has been involved in research and teaching in the area of death and dying. She chairs the highly successful Open University course on death and dying and has been involved in several research projects in end-of-life care. Carol has a practice background in healthcare, midwifery and counselling and is committed to the belief that sociological research should make a difference to the experience of service users and providers.

Susanne Langer is a Senior Research Fellow at the University of Liverpool. A social anthropologist by training her work on death draws on long-standing interests in personhood, bureaucratic technologies, health and illness, and qualitative research. She is co-author of *Suicide: A Sociological Autopsy*, to be published by Palgrave in 2011.

Una MacConville is a sociologist and a Visiting Fellow at the Centre for Death and Society at the University of Bath. Her research interests include palliative care, religion and spirituality with a particular interest in forms of after-life communication. Although a sociologist, Una has a background in archaeology and her interest in the past and in cultural aspects of issues of mortality is reflected in her ongoing work on roadside memorials in Ireland.

Regina McQuillan, FRCPI, is a Palliative Medicine Consultant working at St. Francis Hospice, and Beaumont Hospital in Dublin. She has a particular interest in providing end-of-life care for marginalised groups. She also has an interest in how death and bereavement are dealt with by the public and in the public, including public commemoration of death and media representation.

Susie Page currently works on a freelance basis lecturing at both Oxford and Greenwich Universities in addition to leading on a number of high-profile research and evaluation studies. She has recently been working with selected NHS organisations as they seek to implement the Transforming Community Services policy agenda. She is now leading a work stream for the professional support organisation, Local Partnerships, around capability development within the public sector workforce whilst continuing as a Visiting Fellow at the University of Greenwich.

Mark Powell completed his PhD in Social Anthropology at the Queen's University of Belfast in 2006. His doctoral research considered issues of identity, belonging and community in the context of the Eastern Caribbean. He has since pursued his interest in identity through his work as a researcher on two ESRC-funded projects at the University of Sheffield. The first of these examined the landscape experiences of first-generation migrants living in Sheffield, and the second focused on research into the cultural, social and emotional implications of 'natural burial' in the UK. Mark is currently a Research Associate at Durham University, working on an ESRC-funded research project that considers disaster responses to the post-tsunami situation in Sri Lanka.

David Prendergast is a social anthropologist in the Digital Health group at Intel and a Principal Investigator of the Social Connection Strand of the Dublin based Technology Research for Independent Living (TRIL) Centre. His research focuses on later life-course transitions and utilises ethnographic research to co-design and iteratively develop culturally appropriate assistive technologies for older people. He has authored a number of books and articles on ageing, health, homecare and death. He is the author of *From elder to ancestor: old age, death and inheritance in modern Korea* (Global Oriental, 2005).

Caroline Simone is the Joint Honours Subject Leader of Applied Studies at the University of Derby. She also lectures in the Sociology subject area and delivers their Death and Society module. Caroline's PhD, Bereavement after Parental Suicide: Transcending Chaos and Disorder, focuses on the long-term impact, complexity and management of bereavement loss experiences after a parental suicide, or self-inflicted death. She is also a trustee of a national bereavement organisation, Survivors of Bereavement by Suicide, and a leader for a suicide bereavement self-help support group.

Tony Walter has written and lectured extensively on funerals, public mourning, afterlife beliefs, representations of death in the media, mediators between the living and the dead, the Body Worlds exhibition, pilgrimage, and the concept of spirituality in palliative care. He runs the MSc Death & Society at the University of Bath.

Kate Woodthorpe is a lecturer in Sociology at the Centre for Death and Society in the Department of Social and Policy Sciences at the University of Bath. She has published on the material culture of cemeteries and the experience of undertaking research in places associated with death. Her current research with Carol Komaromy is examining the role of the anatomical pathology technologist in UK hospital mortuaries.

Ken Worpole is a writer and broadcaster, and the author of many books on social policy, landscape and architecture. He was a member of the UK government Urban Green Spaces Task Force, and an Adviser to the Commission for Architecture and the Built Environment (CABE) and the Heritage Lottery Fund. His books include, *Here Comes the Sun: Architecture and Public Space in 20th Century European Culture* (2001), *Last Landscapes: The Architecture of the Cemetery in the West* (2003), and most recently, *Modern Hospice Design: The Architecture of Palliative Care* (2009). He is currently a senior professor in The Cities Institute at London Metropolitan University.

Preface

This collection has its roots in the *Social Aspects of Death, Dying and Bereavement Symposium* held in Sheffield in November 2006. Speakers were invited to address the theme of material culture in their papers and a selection of their work forms the cornerstone of this volume. The collection is the fourth in a series that stems from this long-running symposium; David Clark edited *The Sociology of Death* in 1993; David Field, Jenny Hockey and Neil Small *Death, Gender and Ethnicity* in 1997, and Jenny Hockey, Jeanne Katz and Neil Small *Grief, Mourning and Death Ritual* in 2001.

David Clark, David Field and Nicky James led the first *Social Aspects of Death, Dying and Bereavement Symposium* in 1990, and this successful, annual event has since been hosted by staff at a variety of UK universities. It has stayed true to its original aim to provide a forum for developing new research ideas and continues to attract a mix of postgraduate students, early career researchers, practitioners and established academics in an intellectually challenging, enabling day of plenary papers. The British Sociological Association gave support and recognition to the symposium within the Medical Sociology Sub-Group. In 2009, Kate Woodthorpe at the University of Bath secured its status as a sub-group in its own right and has now taken on the role of its coordinator.

The current expansion of Death Studies and its scope for theoretical, empirical and policy-oriented research owes much to the *Social Aspects of Death, Dying and Bereavement Symposium.* Along with the journal *Mortality*, the international *Social Context of Death, Dying and Disposal Conference* and the Association for the Study of Death and Society (ASDS), the symposium has provided a touchstone for established academics as well as a valuable career development opportunity for postgraduate students, practitioners and early career academics.

1
Materialising Absence

Jenny Hockey, Carol Komaromy and Kate Woodthorpe

How death manifests itself within the western social world is the primary focus of this collection which takes us into hospitals, hospices, residential homes for older adults, funeral directors' premises, coroners' courts, cemeteries and burial grounds, as well as less obviously death-related landscapes such as roadside memorials and informal ash-scattering sites. We conclude with a visit to the more metaphorical terrains within which deities and spirits are imagined to reside, and here we consider beliefs and practices in more traditional as well as western societies.

Many of our contributors, whose work leads in these disparate, death-related directions, are social scientists. They have drawn on literatures within sociology, anthropology and material culture studies, using concepts such as identity, embodiment, personhood, agency and ontological security, as well as models or schema such as *rites of passage* (see Van Gennep, [1909] 1960). However, the collection is also enriched by contributions from historians Jan Bleyen and Clare Gittings; landscape architect Andy Clayden; sociologist and bereavement support practitioner Caroline Simone; theologian and anthropologist Douglas Davies; and policy analyst and architecture/landscape historian Ken Worpole. The collection therefore draws on a range of disciplinary resources, which are reflected in the lively variety of presentational styles it encompasses. Our introduction provides a theoretical overview within which readers can locate particular chapters. To this end, it sets out the range of analytic, sometimes discipline-specific, perspectives that contributors draw upon.

Our title, 'The Matter of Death', is a deliberate pun that signals the collection's concern with both the meaning of death as part of human experience and the materialities associated with its 'management'.

However, along with the matter of death, many chapters raise a related question: what is the matter with death? Sociologists (Giddens, 1991, Bauman, 1992, Mellor, 1993, Seale, 1998) and anthropologists working in western societies (Gorer, 1965, Hockey, 1990, Lawton, 2000) have consistently critiqued twentieth- and twenty-first-century western death-related attitudes and practices, arguing that these have led to it becoming 'taboo', 'denied' or 'sequestered'. In part this change can be seen as a result of nineteenth- and twentieth-century improvements in public health and medicine, many of which have a basis in western economic growth and stability; death happens less often among children and young adults (Jupp and Walter, 1999). Whether its relative unfamiliarity has made it harder to bear, and so made its 'denial' necessary, or whether western economic stability has facilitated specialist services which intensify the consequences of its relative rarity by sequestering it from the workaday world, is difficult to ascertain. Whichever explanation is adopted, westerners, somehow, are seen to be getting it wrong (Hockey, 1996).

These arguments will be addressed throughout the volume, and powerful empirical evidence of 'getting it wrong' characterises many chapters. Evidence is not, however, left to speak for itself, and the following section of this chapter details the theoretical resources that have been drawn upon to explain what, how and why dying and loss have come to be managed in particular ways. Furthermore, the locatedness of our topic needs to be noted: western death in the twentieth and twenty-first centuries. Thus, while both the certainty of death and the uncertainty of its timing are shared universally by all human beings, making sense of how particular people think, feel and act in relation to it requires that their responses be contextualised, culturally and historically. To what extent, we ask, are responses to death amenable to change? Hence our work not only explains death-related policy and practice but also informs it with a view to creating change, as Chapter 15, 'Recovering Presence', underlines.

Making sense of death

As many chapters in this volume demonstrate, a social constructionist perspective has allowed researchers to identify the models, paradigms or classificatory frameworks which operate as organising principles when human beings encounter death. What people understand, believe and do is likely to reflect broader historical and culturally specific models of the world in some way. Anthropologists, in particular, have concerned themselves with the ways in which people create order and indeed knowledge of themselves and their environments.

The notion of classification thus assumes the possibility that the members of particular societies divide up the world into categories – for example, this is burial space, this is work space, this is home space; these people are my family, these are my community, these are strangers and so on. While these particular divisions may not be evident in all societies, one of this volume's key areas of theoretical interest are differences between the ways in which categories such as these are used, depending on a particular society. For example, Chapters 4 and 6 present data from residential homes for older adults and coroners' courts, respectively, and describe how people differentiate between different kinds of death, citing Bloch and Parry's (1982), Catedra's (1992) and Bradbury's (1996) ethnographic accounts of 'good' and 'bad' deaths and the practices undertaken in order to achieve a 'good' death. Prior's (1989) account of the social organisation of death in Northern Ireland provides further evidence of the way in which a society can categorise deaths.

In addition, divisions between categories, as well as who or what goes into them, have been a focus for considerable work – for example, this is the boundary between my land and theirs; this is the boundary between time for work and time for religious activity; this is the boundary between being alive and being dead. However, this description implies the possibility of imposing a fixed classificatory grid onto some kind of freeze-frame world – and, as work on death in particular reminds us, human beings and their environments are in constant flux, moving between categories such as childhood, adulthood, later life and death; between winter, spring, summer and autumn.

Early anthropologists such as Hertz ([1907] 1960) and van Gennep ([1909] 1960) addressed what they saw as a potential tension between ideas and materialities – one that put systems for ordering the world under threat – by investigating rituals of transition such as healing rites or puberty, marriage and death rites. Taken together, these theorists provided the basis for a model or schema of how transitions between categories are achieved. Hertz ([1907] 1960) described examples of Southeast Asian death ritual that involved two rites separated by a period when everyday life was put in abeyance, while van Gennep ([1909] 1960) went on to develop descriptions of comparable processes in his rites of passage schema. Van Gennep saw the in-between time that linked separate categories as a threshold, or limen. Safe transition across this threshold could be achieved by suspending customary rules and roles and engaging in ritual practices that undid membership of previous categories, enabling people to assume new identities, such as 'adult', 'widowed' or 'ancestor'. Being betwixt and between statuses was therefore to be in a state of liminality. As Chapters 4 and 7 explain, van Gennep ([1909] 1960) believed these ritual practices served important psychological, sociological and symbolic functions by ensuring the orderly progression between categories.

Van Gennep's work on rites of passage was originally published in French in 1909, but when it was translated into English in 1960 it came to be developed theoretically, as this collection demonstrates. For example, Douglas (1966) focused on those things, places and people that failed to fit into classificatory schemes, instead straddling boundaries between categories. A 'bad' death, therefore, might be one which did not happen at the right time, or in the right place. It becomes, as Douglas (1966) said, matter out of place. She used this concept to explain that commonsense ideas about pollution or dirt were not necessarily about biomedical dangers, instead referring to that which 'messed up' a classificatory system. Chapter 6 describes deaths of this kind being subjected to the proceedings of the coroner's court in order to establish in which category they might be placed. As long as a death resides awkwardly within a society's classificatory system, it remains ambiguous, fitting neither here nor there. Regardless of whether such a death remains forever impossible to categorise, or is simply in a state of transition or liminality, Douglas (1966) argues that it becomes invested with power and is potentially dangerous, requiring (ritual or symbolic) management. Chapter 11 describes burial on private land, showing how disposing of a body outside customary public spaces can evoke concerns about its symbolic 'safety'. Such concerns were exemplified when the bereaved relatives' more distant kin sought to re-order disposal in a farmer's field by bringing conventional grave markers. Chapter 9's account of the often slow decision-making process undertaken by someone removing ashes from a crematorium for independent disposal also points towards the implicit risk that choosing the 'wrong' place might result in ashes becoming 'matter out of place'.

Thus far, we have described somewhat abstract classificatory systems. However, anthropologists such as Fernandez (1972) and Turner (1975) argued that liminality *makes* as well as marks transitions between categories. Indeed Turner (1975) argues that liminality is the creative space out of which a classificatory system emerges. This raises the question of how this is achieved and indeed what it is that holds a particular classificatory system in place. Chapter 2, for example, describes expensive and sometimes violent medical practices widely undertaken to return people from the category 'dying' to a stable position within the category 'living', ones which routinely fail to achieve this. How can we explain the persistence of these practices? Social theorists have answered this question in a variety of ways. Sociologist Foucault (1973) argues that systems of ideas – or, as the anthropologists we have discussed might put it, classificatory schema – become powerful by virtue of their dissemination throughout society. Describing such systems as discourses, he identifies their scope for dispersal, using the metaphor of the circulation of blood throughout the human body. This 'capilliary action', he suggests, ensures that the power of a particular regime of ideas takes effect locally – in the Accident and Emergency (A&E) department for example, where a high-status western medical discourse provides a powerful rationale for apparently unproductive practices, as Chapter 2 demonstrates. Chapter 8 provides a fascinating account of what happens when one set of ideas, or discourses, is compounded by another. In this case, health and safety regulations which underpin the practice of removing memorial items from graves in a cemetery intersect with lay interpretations of bereavement theories which discourage open-ended, materially grounded relationships with people who are dead.

Anthropology is helpful here as another source of theoretical understanding as to how a particular set of ideas is made to stick. For example, it has informed work on metaphor that shows how ideas from a domain life that has a familiar, taken-for-granted status can be used to create 'knowledge' about one that is more abstract, mysterious or frightening (for example, Fernandez, 1972, Lakoff and Johnson, 1980). As Chapter 5 shows, the 'bad' death of a baby around the time of its birth confounds the contemporary western sequencing of boundaries between birth, life and death. In the earlier, more traditional societies that van Gennep ([1909] 1960) describes, stillbirth and neonatal death appear not to have been constructed as disruptive, such babies instead being treated as non-persons that had yet to enter any kind of classificatory system. By contrast, for contemporary western parents, particularly those experiencing difficulty conceiving a child, the newly fertilised egg can be felt to possess a social identity as a being on its way to personhood (Hockey and Draper, 2005). Yet, with at best limited tangible substance as a living daughter or son, that person-to-be has an uncertain, mysterious status that confounds attempts to position him/her/it within social space. Nonetheless, through the metaphoric use of entities and objects drawn from more familiar domains, that child takes on more enduring substance. Thus, Chapter 5 describes a man bereaved of his newborn godson addressing 'him' at his funeral in a way that acknowledges his elusiveness yet still grants him a material form: 'Wherever you are, you little gnome, we wish you all the best.'

Utterances of this kind have a degree of symbolic efficacy. Other family members may find such a statement memorable, retaining and re-using images evoked during a rite of passage such as a funeral. However, this goes only part of the way towards explaining how (ritual) practices can make things happen. Anthropologist Turner (1975) provides a valuable theoretical perspective here, showing how metaphors acquire their symbolic efficacy through being embodied in tangible entities where they then inform practices within material contexts, so making the abstract knowable and, potentially, sufferable (see also Fernandez, 1972, Hallam and Hockey, 2001). Chapter 12, for example, describes the metaphors used by two adults bereaved of a parent by suicide during childhood to describe their therapeutic attempts to reduce the emotional costs of these losses. Using the spatial metaphors of 'walking the walk', 'sentimental journey' and 'the re-enactment' did not simply give figurative shape to a narrative; the persuasive or rhetorical power of these images derived not only from their familiarity within other domains of life but also from their ritual, embodied enactment. Both individuals went to the sites of their parents' deaths, places they had avoided as unassimilable into any system for ordering the past. Remaining liminal, the sites of these violent deaths were polluted, dangerous and above all powerful (Douglas, 1966). Materially grounded, spatialised metaphors used during the research interview described in Chapter 12 took on the power to construct reality (Lakoff and Johnson, 1980) once they were acted upon, thereby becoming tangible vehicles that allowed loss to be made sense of and childhood memories to be (re)generated.

Material culture studies

The title of the previous section – 'Making sense of death' – quite appropriately signals its predominantly cognitive focus. It began with the abstraction of classificatory schema or discourses and then addressed the question of how this kind of system acquired persuasive power and came to be materialised in particular social arrangements and culturally specific practices. The embodied nature of human life and the materialities of both our bodies and their surroundings are indeed addressed within theoretical perspectives of the kind so far laid out. However, the materially grounded, phenomenological perspective which has inspired this collection allows us to theorise more precisely the ways in which human beings manage their own mortality.

Epistemologically, the phenomenological approach which is central to much of material culture studies argues that we know the world only as an outcome of our embodied engagement with it. We have no other place to start from. Out of the indeterminacy of space and matter, places and objects emerge through our 'being-in-the world' (Csordas, 1994). The imagination – which might appear to be disembodied – is nonetheless fuelled by experience which has the body as its site of knowing.

Leder (1990) elaborates this point in his work on the absent body which in good health effaces itself, so enabling embodied perception of the surrounding physical and social world. When we change gear in a car or manipulate a computer mouse, it is the road ahead or the words on a screen that engage us, not pressure on the skin of our hands or the flexing of its delicate musculature. In the case of an illness such as repetitive strain injury, however, the urgent demands of painful hands can inhibit perception of the oncoming lorry or the route through a website. This leads us into the anthropologist Ingold's (2000) sustained critique of the idea that the environment passively receives a set of cognitively derived meanings that human beings impose upon it, an approach which the earlier anthropological theorising presented above exemplifies. Using the concept of a dwelling perspective, he argues that just as being incarnate makes thought and experience from a disembodied perspective impossible, so dwelling in the world excludes any scope for generating a 'detached' or purely cognitive understanding of that world.

Tilley's (2002) discussion of the relationship between ideas and material culture elaborates the value of this perspective for investigating responses to death. Merleau Ponty argued that 'the world is the homeland of our thoughts' (cited in Ingold, 2000: 186), and Tilley's account of the metaphoric nature of human thought develops the discussion introduced above. While Levi-Strauss highlighted the metaphoric resonance of the materialities of the house, showing how forms of social organisation were mapped in its spatial layout and functioning (see Buchli, 2002), Tilley questions the cognitive focus of this approach, describing metaphoric thought as the outcome of 'the sensuous and carnal capacities of the human body' (2002: 24). Thus, embodied experiences of the 'stuff' of everyday life are, as already argued, pressed into metaphoric service to help make sense of more mysterious or abstract domains: 'when we link things metaphorically we recognise similarity in difference, we think one thing in terms of the attributes of another' (Tilley, 2002: 24). So, for example, similarities have been recognised between climbing into bed, relaxing into unconsciousness and awakening refreshed 8 hours later, and the more permanent cessation of bodily functioning at death. Just as the sleeper's dreamworld is relatively resistant to exploration, the subjectivities of the dead remain a matter of speculation or faith. Neither their extinction nor their survival can be proven. The recognisable similarities between these very different aspects of embodied life have resourced not only conceptions of death within Christian cultures that posit a sequencing of death, resurrection and eternal life; they also inform western practices such as quietness or silence around the time of death, and the bed-like design of the traditional grave where sleeping meshes with metaphors drawn from the body and its parts – and the deceased person – is 'laid to rest' below their 'head' stone.

The privileging of sleep as a metaphor for death is, however, one among the many materially grounded experiences which resource responses to death. Discussion of the geographies of spirits in Chapter 14 indicates that making a journey, such as a pilgrimage, is another such experience, with its elements of transition from one place and embodied state to another, of potential exposure to danger, and the dimension of movement towards a destination of some kind. Again, similarities between journeying and the very different domain of death reside in the notions of change, danger and transformation which both aspects of human experience can evidence. For example, as Chapter 3 details, the decision to set up 'hospices' for the provision of palliative care for people who are dying has metaphoric resonance with the hospices that traditionally acted as way stations for pilgrims journeying towards a sacred destination. This thinking-through-dwelling which lies at the heart of a material culture perspective thus provides a framework for understanding how any one way of engaging with death comes to hold sway, given the diversity of human thought and experience.

As this collection shows, therefore, the materialities of place, things and bodies that are lived through in practice; the social hierarchies that rest upon and within the dispositions of institutions; and the affordances of the body itself, its material scope or capacity (Gibson, 1979), can all be recruited to particular systems of power. That said, in the constraints and limitations of an ultimately intractable material world, including the materiality of the body itself, we find resistance to the agency of human beings – as well as the resources through which the members of one social category might resist the power of another. For example, among older adults living in residential care, bodily incapacities such as loss of mobility or continence can become sites of resistance, residents subverting the smooth unfolding of institutional timetables as they fall over or have 'accidents' (Hockey, 1990).

If a social constructionist perspective not only privileges systems of cognitive rather than embodied knowledge but also understands human experience as the outcome of their imposition upon an external world, a phenomenological perspective treats human beings as integral to, rather than separate from, the material world that surrounds them. This position leads to a rethinking of theories of agency and the notion that it might inhere in inanimate as well as animate entities. Traditional social science models bind agency with intentionality, arguing that true agency is only manifested in human beings who have the capacity to act upon their own decision-making processes. Yet, as was evident even when a more cognitive model of the social world was being discussed earlier in this section, that which is either temporally or permanently betwixt and between the categories through which the members of a particular society understand the world acquires power by virtue of its disruptive status as 'matter out of place' (Douglas, 1966). However, once we take on the idea that knowledge of the world emerges out of embodied engagement within it, we need a more comprehensive understanding of agency.

If, following Latour (1993), we adopt a model of agency as the capacity to have effects, eschewing any separation between human beings and their environments, we are likely to discover these effects to be far from uni-directional. Rather, effect happens within networks of human beings, inanimate objects and their surroundings. Appreciating agency in terms of effect therefore allows us to recognise human beings and their environments as co-producers of meaning, a stance that leads us to consider the agency of not only individuals but also the hospitals, coroners' courts, funeral homes, cemeteries and domestic environments in which they die and grieve. Gell (1998) adheres to a similar model but chooses to make some differentiation between the agency of human beings and inanimate objects and matter, retaining a more traditional link between agency and intentionality. That said, he attributes the non-human world with secondary agency and is concerned with the ways in which objects can become extensions of human beings, acquiring agency as a result of human intentionality. As such, objects constitute potential resources for distributed personhood, a model of the individual which transcends the boundaries of the body (Gell, 1998). An example of this would be the ways in which an individual's personal belongings, their clothing, shoes, watch and jewellery, not only constitute part of their personhood, but also, once they have died, assume enhanced agency as residual fragments which can stand for that individual in their entirety. Chapter 6 draws on this model to show its reverse operation in the coroner's court when human beings convert their primary agency as individuals into a form of secondary agency, akin to that of an object. Through the style and content of their witness statements they can take on the status of what Mauss (1987) has called personas; that is, they act as representatives of a particular social role, with consequences for their degree of individual responsibility for their acts. As Chapter 6 notes, within such a process aspects of individuals and personas are intertwined, an oscillation that lends the inquest some of its potency.

The case of the body

Having concluded a theoretical journey which encompassed a wide range of approaches to making sense – or making sufferable – the matter of death, we bring these approaches to bear on the body itself as a key and constant player within the death-related trajectories explored within this collection. In so doing we consider in more detail their value as resources for understanding death-related experience. While the spatial or institutional sequestration of people touched by loss has been a focus within the social sciences (Giddens, 1991, Bauman, 1992, Mellor, 1993, Seale, 1998), radical physiological changes to the body might seem to be the more pressing 'matter of death'. Someone 'like us', someone we might have loved, whose body we knew intimately, as lover, parent or carer, who was warm and animated, is now cold and inanimate. The body that sustained itself, whether for a few hours or the lifetime of a long marriage, is now unsustainable. Without our intervention, it will undergo the profound changes of deterioration and decomposition, eventually revealing the scaffolding of bone beneath the flesh. Crace (1999) structures an entire novel around this process in his account of events leading up to a couple's murdered bodies lying undiscovered on sand dunes for 6 days.

How people react to the apparently intractable corpse reflects their relationship with the individual who has died and is therefore a socially constructed, rather than 'given', response. And yet the materiality of the dead body can evoke powerful emotions, over and above grief that someone's life has ended. As an entity it seems to demand a response, as Verdery (1999) argues in her exploration of the corpse as a political resource. In her view, its symbolic efficacy derives from its materiality as a concrete object which is amenable to sensory engagement: touch, sight and smell. However, she is not suggesting that the corpse evokes a fixed set of responses. Rather, as the material relic of a particular individual, the dead body gives the powerful appearance of having a single meaning, a quality which lends authenticity to whatever meanings are imposed upon it. Importantly, it does not talk back.

In acknowledging the importance of the dead body's materiality and considering how this contributes to its symbolic efficacy, Verdery's (1999) work expands upon the cognitive focus that characterises many social constructionist approaches. In terms of the living body, there are arguments to suggest that social constructionist accounts are somewhat limited in relation to the body we *are* – our embodied selves – in that theorising tends to reflect a position that is external to the body. Thus, the sociology and anthropology of the body often treats it as an entity that we *have* (Turner, 1992, James and Hockey, 2007). However, the dead body complicates this distinction. For example, in contexts such as the mortuary where the deceased person is unknown to those handling their body, the body they *had* can be said to have survived the body they *were*. The subject body has become the object body. Indeed, social identity can be effaced while the corporeal body is still alive; older adults with dementia have been described as a corpse the funeral director has omitted to collect (Miller, 1990, cited in Hallam *et al.*, 1999: 3). Conversely, in Japan, 'death' does not coincide with the cessation of the heartbeat (Lock, 2002), instead the individual remains socially alive, their 'death' being the gradual outcome of family-based processes of transition. Chapter 7 provides an excellent account of how the social identity and the dead body are not only merged but also separated in the course of different western death rituals, UK funeral directors sequestering the corpse as the site of exclusively private social identification, allowing it to be more publically reconciled with the social identity of the deceased person only later, when coffined and surrounded by floral tributes. In the United States, the reverse trajectory is enacted, the corpse initially being made to stand for the social identity of the deceased person in elaborate rituals which bring family and friends together with the spot lit, highly decorated body. Only after its journey to disposal begins does the corporeal body begin to diverge from the social identity of the dead person.

Always in flux, then, the human body's ultimate resistance to efforts to retain it as the immediate, tangible site of social identity represents one of its affordances (Gibson, 1979); its material scope or capacity.

Whether alive, dying or dead, human bodies have corporeality and exist in time and space. As Frank (1991) argues, they are produced by bodies and comprise flesh that is formed in the womb, transfigured in its life, dies and then decomposes. This apparently seamless physiological process can be seen as the indeterminate 'matter' which can be objectified into differentiated categories such as 'newborn', 'adult', 'frail' and 'dead'. Seeking to explain how these categories are made to 'stick', Frank points towards an equilateral triangle made up of the body, institutions such as the church and discourses such as medicine. Ideas, social systems for managing human growth and deterioration, and the body itself, when combined, result in sets of 'given' ideas and practices.

In that corporeal processes of growth and decay are apparently seamless they are, therefore, both amenable and resistant to human intervention. Thus, the social categories through which we 'know' the body – and its transitions through the life course – cannot simply be imposed upon it from without. Instead, as Frank (1991) argues, corporeality and the space that the body consumes reveal its ultimate resistance to imposed control and management and, within some societies, its status as a problematic entity. This lack of predictability is a feature of Chapters 2, 4 and 6 where the body – or testimonies relating to the body – is a site of activity geared towards making the body perform or take shape in a particular way.

Such activities can be profoundly successful, even though the body might retain scope for resistance; for example, Mauss's ([1934] 1973) concept of bodily techniques shows even walking to be a skill learned in a particular social fashion; Goffman's (1968) work on total institutions demonstrates their capacity to control the bodies they house; and Bourdieu's (1977) discussion of pedagogy shows conformity being exacted through the body. When dead, the body might seem an unlikely site for any further resistance. Yet, as Chapters 5 and 7–11 all show, the disposal of dead bodies and the places in which the bodies reside require careful management and control. Indeed, even after their disposal, bodies still require some form of management as well as retaining the capacity to evoke powerful responses. As Synott argues, 'attitudes to the body persist after death' and the body remains the 'powerful symbol of the self' (1993: 33).

While these perspectives emphasise the status of the dead body as a site of agency, pollution and danger, the body nonetheless has considerable scope to be surveilled, controlled, produced and defined, and during the latter part of the twentieth century this became the focus of many disciplines beyond a biomedical sphere. The works of theorists such as Giddens, Foucault, Turner and Frank, for example, are relevant for this collection, because while the body work that takes place around the dying and dead body is an indication of the professionalisation of death and dying, these theorists provide explanations for the way in which professionalised work takes place. In other words, the chapters here explore what else is taking place beyond that which is observable on the surface. Dying, dead and disposed of bodies thus provide an emergent theme in this collection on several levels: at the level of regulation, control, and resistance and as a container of death.

The body as a site of regulation

In focusing her ideas on old bodies, Tulle-Winton (2000) argues that an avoidance of the consequences of the relatively intractable indeterminacy of ageing bodies – for example, their capacity to act as a threat to western norms of identity and the valorisation of youth – has resulted in an extension of middle age and the marginalisation of those who cannot escape old age and are likely to be old and frail. While this collection is not focused on the ageing body per se, the body that is close to death carries similar problems which are also in part regulated through the 'othering' of those bodies which younger adults will, nonetheless, one day become. The concept of stigmatisation has been a focus for sociological consideration for over 50 years, as a result of the work of Goffman (1963a), and here helps make sense of the activity focused in and around body maintenance and body work. Tulle-Winton (2000) uses the example of ageing bodies to argue that if bodies cannot be self-regulated, that is, subjected to the embodied intervention of the person who has that body, then they become the responsibility of state agencies which use measures to evaluate and 'treat' their dependency and loss of function.

Foucault's (1973) focus on the body as the outcome of discourse has a contribution to make to our understanding of the body we have, even though his cognitivist neglect of embodied processes of engagement with the body as the site of 'knowing' has been criticised from phenomenological perspectives (see Csordas, 1994, Ingold, 2000). Foucault (1973) was particularly concerned with the self-disciplining of the body as a way of controlling and making it predictable through regimentation. Lupton (2004) has applied Foucault's ideas about the disciplined body to the notion of the new public health in which the body is regulated through regimes of discipline. Elsewhere, Hughes (2000) argues that health promotion offers people freedom from medicine, even though the regimes of body discipline impose just as rigid a form of discipline and control as medicine. In contemporary western societies, disease is not just something to be overcome; it is a failure of health maintenance (Hughes, 2000). Similarly, Kellehear (2007) argues that death and dying are used in public health discourse as threats of the consequences of failure in the regulation and maintenance of healthy bodies. But what happens to the body that has failed to survive or keep itself alive? While disciplined and healthy bodies are only indirectly susceptible to medical control, dying bodies are more overtly submitted to interventionist institutional and medical control, as Chapter 2 illustrates. However, the idea that the body is a site of focus for medical discourse only offers a partial explanation of what is happening in death and dying. It is important to recognise the bodies we *are* as negotiated territories, even in their most vulnerable states.

The body as a site of resistance

The notion that power flows in one direction only has been hotly contested; for example, Foucault's (1973) notion that power is not in a direct relationship with a higher entity or institution, but instead diffused through discourses and regimes of truth, allows for a different understanding of control and resistance. If it is the case – as Foucault argues – that identities are produced at the intersections of the regimes – then these can also be understood as the sites where resistance takes place. For example, as Hughes (2000) argues, with respect to medical discourses people in western societies are living in a post-professional age where the doctor–patient relationship that was founded on lay people with problems and experts with solutions has been transformed into one where a medical knowledge is consumed alongside mediatised health information such as NHS Direct and illness-specific self-help groups.

However, dominant discourses that produce a fragile (dying) body might seem to admit fewer alternative regimes of truth and scope for resistance therefore might seem considerably reduced. Yet as Chapters 2 and 4 highlight, bodies do not act as they ought to in response even to powerful exertions of physical control. Older people who die suddenly, regardless of the extent of institutional order and surveillance, die 'without permission', thereby creating the potential for chaos. Also, as Chapter 2 records, hospital in-patients who refuse to respond to the violence of CPR, in their resistance to the best that medical technology can offer, become medical failures rather than evidence of the failure of medicine.

The body as a container of death

Once the body approaches death, it comes to represent the site of the western boundary between life and death and, as a marginal entity disrupts the relationship between the lived and the object body, the body we are and the body we have. Representing the limit or margins of life, it acquires a potent, categorically ambiguous materiality. As such, its agency is likely to be managed through concealment which ultimately terminates in disposal. For example, Lawton's (2000) empirical work on the relationship between identity and dying based in a hospice revealed an unexpected dimension that she called 'dirty' dying. For those people whose body's boundaries could not be controlled and for whom smelly and leaking fluids breached the body's boundaries, corporeality became their dominant expression. According to Lawton, dying people withdrew into their bodies, not wanting to engage socially with staff or visitors. In turn, family members were likely to withdraw and carers would isolate these dying people within single rooms in an attempt to conceal visible evidence of leakage, even if its smell continued to pervade. Thus the corporeal boundaries of the body, in proving resistant to medical management, were supplanted by the bricks-and-mortar material boundary of a single room within a hospice. The activity that takes place in managing such bodies is also evident in Chapter 4 in the way that the dying bodies of older people, as they unravel their way to death, are concealed in the private spaces of care homes. This mirroring of the institutional response to the disruptive bodies close to the moment of death seems to be overdetermined and arguably might be a last-minute attempt to contain death, if not within the disintegrating boundaries of the corporeal body, then within the material boundaries of an institution. Only after the demise of the body, and its subsequent cleaning, plugging and wrapping, can it be restored, materially, to its status as a (relatively) safe container of death. Chapter 6, however, shows that even after death, the body can be literally and metaphorically 're-opened' in post-mortem investigations designed to reveal its secrets as part of the evidence for the coroner's court.

Once the body ceases to be of interest or relevance for a medical profession committed to sustaining life, it becomes matter that is seen to need imminent disposal. Reiterating the work of Douglas (1966), discussed above, the deceased body and its contents can become 'matter out of place', materialities that are categorically ambiguous, being resistant to felt needs for ordering until their disposal. Thus, for Douglas (1966), things are not of themselves intrinsically 'dirty' or 'polluted' – their status is indeterminate; however their placing within social time and space dictates the extent to which they become dirt or matter out of place. The body in coronary care that has failed to respond to the attempts to resuscitate it and dead bodies in spaces preserved for living bodies in care homes thus create discomfort that requires forms of management. It follows that the dead body, as a container of death, is therefore not just potentially polluting; it also symbolises the fragility of the boundary between life and death. The corpse is a memento mori that reminds living people that while death is their only certainty, its timing remains uncertain. The activity that follows death can therefore be seen as part of the iterative process through which individual constructs are mediated by embodied engagement with the materialities of disposal and loss, a process which both draws upon and in turn shapes cultural values and categories. Thus, activities surrounding bodily death can be seen to have ritual dimensions, in that they are implicitly oriented towards restoring symbolic order. As such, they can constitute more than simply instrumental necessities determined by professional and medical discourses on the dead body.

Of course, once dead, there is the matter of the body's disposal and, drawing on the anthropological literature referred to earlier in this chapter, the 're-ordering' of the social world. Examples of attempts to re-order the world in this volume include Chapter 6's analysis of the process of publicly accounting for an individual's death in an inquest and the collation of coroners' files that come to represent a life once lived, so enabling social order to be restored, and Chapter 5's description of the creation/destruction of an embodied identity/personhood when embodied life never got started. Together these chapters reveal the way in which different environments and the organisational/institutional practices that take place within them (the assembling of the inquest file, the management of the stillborn baby's body in the hospital ward) enable or undermine the agency of the body, and thus restore order, of the person-that-once-was.

Absent bodies

At the heart of this collection is the matter of death: the stuff, the spaces, the practices which surround the ending of embodied life, the disposal of the body, the mourning and the memorialisation which then ensue. Yet a curious paradox lies at the heart of our endeavour. With its focus on contemporary western society, the collection is exploring a society where death is associated with invisibility and disappearance: the confinement of the dying body to the hospital ward or the private home, the removal of the corpse to the mortuary or funeral parlour, the social isolation of bereaved people and the emotionally muted funeral. Walter (2008) argues that mourning and memorialisation have become more public, whether in the form of items left on a grave, roadside memorials or public benches bearing plaques dedicated to people who have died. Yet Chapters 9–11 show that the disposal of the body seems to involve its disappearance in that its ash residue may be invisibly scattered or interred at sites which bear no visible memorial; natural burial involves minimal grave-marking and in many cases eventual disappearance of any accessible grave site; and burial on private land is likely to conceal the presence of the dead person as a member of any kind of collectivity, as manifested in a cemetery.

Why then are we making material culture our focus when death in contemporary western societies seems to involve the disappearance of not only the human body but also its site of disposal? The discussion of stillbirth in Chapter 5 provides a fine-grained analysis of how absence, in practice, intersects with presence. In other words, the invisibility of someone who has died – perhaps most tellingly when they had no previous embodiment lodged in the memories of their survivors – can work to powerfully evoke their presence. Thus objects which stand in for the child who has died, the cradle, the doll and the mushroom, all serve a figurative purpose in making present the absent baby. Disposal practices themselves similarly evoke that which they render invisible. Since the early nineteenth century in the United Kingdom and Europe, efforts have been made to confine the dead to the peripheries of communities (Hallam and Hockey, 2001, Vanderstraeten, 2009). Reflecting a sequestration of death more generally, these walled cities of the dead (Warner, 1959) segregated the living from the buried bodies that resided there. However, as populations have grown and suburbs expanded, many of these spaces are now in the midst of busy communities. With burial space at a premium, Chapters 8–10 describe a trend towards the return of the dead to spaces inhabited by the living, their meaningful mementoes left at the graveside, their ashes ground into the interiors of pubs and their burial grounds potentially bounded by the same hedges that mark the edges of similar fields. This penetration necessarily blurs the boundaries between spaces set aside for the dead and for the living.

This type of disposal raises questions about what happens when the dead are entirely absent in any bodily form. Questioned elsewhere, for example, in how the personal effects of the dead should be returned to newly bereaved individuals when a disaster obliterated their owner's body (Easthope, 2009), or the way in which the US Government represents the flag-draped coffins carrying the bodies of fallen soldiers returning from Iraq and Afghanistan in the media (Casper and Moore, 2009), in this volume Chapters 12–14 consider this absence alongside the merging of the memory of someone lost with place (Hallam and Hockey, 2001). Utilising three distinct bodies of literature – psycho/therapeutic, socio-political and theological – these chapters demonstrate further the penetration of the absent dead and their material remnants in everyday, mundane spaces. Death, as these chapters show, cannot be contained within spaces specifically and conveniently 'set aside'.

The role of 'materialities' in death studies

In this collection, the contested or fragile nature of the belief systems and discourses or schema through which death is managed occupy a central position. Drawing on the theoretical perspectives introduced in this chapter, we make embodied experiences of times and places surrounding death our focus in order to develop a nuanced understanding of how particular materialities can mobilise emotion, commitment and action. In this way, the collection contributes to theoretical debates about evidence for a diversity of responses to human mortality, as opposed to arguments that death-induced physical and emotional pain has some kind of universal, autonomous status. Is dying and loss 'what we make of it'? Or should we give more weight – or agency – to the materially grounded nature of human life and death? The collection addresses an apparent impasse between social constructionist arguments that human beings inhabit a world that emerges out of their interpretations, that people see and experience their environment in culturally specific ways – and everyday experiences of the immediacy of our bodily, emotional and sensory responses to that world. As such, the collection speaks to the death studies literature by integrating its focus on materialities and technologies – end of life care, funeral directors' treatment of the corpse, mausoleum and crematoria architecture – with its concern with conceptions and emotions – resurrection and reincarnation (Davies, 2005), 'good' and 'bad' death (Bloch and Parry, 1982; Bradbury, 1996), the grief 'process' (Klass *et al.*, 1996; Walter, 1999).

2
Never Say Die: CPR in Hospital Space

Susie Page

> In western societies biomedicine has occupied a niche in modern life in which its hegemony regarding healing seems to be taken for granted. This hegemony is based on its increasingly high-tech assisted and evidence-based efficacy, on the dominant 'naturalist' and 'positivist' assumptions that bring to biomedicine the air of success associated with the natural sciences, and on the spirit of the Enlightenment that frames the historical development of scientific medicine.
>
> (Lázár and Jonannenessen, 2006: 189)

Introduction

Hospitals and all the high-tech equipment they contain, at least in western societies, are a taken-for-granted part of the geographical and cultural landscape in the lives of the majority of people, and despite the ever-increasing commitment to facilitating 'a good death' in the home situation,[1] it is still the place where most people die.

For many of these people the last active medical intervention that they undergo is cardiopulmonary resuscitation (CPR). CPR is a common emergency intervention undertaken when a person suffers a catastrophic clinical event[2] which places its victim at the boundaries between life and death. If appropriate action is not taken, death will occur. Even with such action, death remains the most common outcome.

Consequently, hospitals are 'geared up' for this event and part of the response means there are spaces within most hospitals dedicated to this procedure, for example, the resuscitation bays of Accident and Emergency (A&E) departments.[3] Equally, the equipment for these events is located not only in resuscitation areas but on every ward and in out-patient clinics and other departments. Thus, virtually any area of the hospital can be converted into a resuscitation space and, as such, these sites are pervasive testimony to biomedicine's gatekeeping role at the life/death boundary.

Despite this and the recent move by *some* medical teams to allow (with discretion) close family or friends to be present during a resuscitation event, the opportunities for anybody over and above the attendant team to witness 'in real life' as opposed to televised drama depictions (see Davey, 2001) what occurs in these spaces are very limited. In general, resuscitation spaces *in action* are, to draw on Goffman (1963b), 'back spaces' inhabited only by 'insiders'; those ascribed particular privileges to be there (see also, Fox, 1997). The drama that unfolds therein is thus primarily played out before a pre-determined and select audience.

The ethnographic study from which this chapter is drawn sought to explore how it is that a predominately unsuccessful medical intervention, CPR, designed originally only to save *'hearts and brains too good to die'*,[4] has become so embedded in western health care practice that it is now a routine response to most deaths.

The intention here is to draw on some of the data from this project to reflect on the space and place in which the study was conducted and on the material artefacts associated with this particular medical practice. Drawing on the thoughts of O'Toole and Were (2008), place is seen as the nexus of things and space within a given boundary – in this case, the place being the hospital. Space includes the exterior and interior spatial arrangements that make up the world and includes quantities, qualities and relationships such as distance and juxtaposition. It is imbued with cultural and political implications. Material culture refers to the corporeal, tangible object[s] constructed by humans (O'Toole and Were, 2008: 617–618).

In this work the spaces of interest are the resuscitation bay of the hospital A&E department and the bedside in a coronary care unit. The material culture of note is the 'tools of the trade' associated with life-saving and CPR which include things like defibrillators and cardiac monitoring equipment as well as drugs, intravenous equipment and devices for maintaining a patient's airway.[5] As indicated, these are concerned primarily with averting death and are part and parcel of the routine and common medically mediated deaths that occur in hospital (and elsewhere of course) reframed, paradoxically, within the vocabulary of saving life. Bodies, dead or alive, move in and out of these 'medical' spaces and often either use or are subjected to the use of the artefacts mentioned.

The medicalisation of death

Death has always and everywhere been a potentially traumatic event and/or caused an interruption to the flow of social life; many different discourses on death exist, for example, the social, emotional and public. The medical discourse, related to those deaths discussed in this chapter, is thus just one of many, but given the context of this study (western acute hospital environments), medicine is the dominant discourse, and in this case the drama around death is predominantly a medical drama (Lupton, 1994). Indeed, as evident from the data reported here, the dying process and the outcome (the death, the corpse) are engaged with, at least in the initial stages, via the beliefs, practices and symbolism of medicine. Medicine and medical artefacts, not religion or myth, are the mediators of death and the force that creates and stabilises the meaning of death. That is to say, the aetiology of explanation for and visual representation of death is provided in essentially biological (scientific/rational/irrefutable) terms, and it is these medical terms that confer meaning on the death. In this context, the 'language of death' is the language of medicine. Congruent with this is the frequent use of CPR at the point of death, even for those who would seek an alternative dying trajectory or for those who have died quietly in their own homes and are then brought to hospital via the emergency services.

Death can be a difficult transition to manage and this certainly appears to be the case for the action and cure-orientated medical model in use in the hospital setting. Seale (1998: 53) and Hockey (2001: 193) remind us that this problem is addressed via the active management of many aspects of death by experts. Seale asserts 'the medical procedures that accompany death have both a technical rationale and a ritual aspect, in that they frame and contain experience and place individual deaths in the context of a progressive, scientific, rational system designed to generate hopes for a better world' (1998: 78); and as Lock suggests, 'we consistently deflect debate around death and transform it into the vocabulary of saving lives' (1996: 596). CPR is a wonderful example of exactly this.

The concept of medicalisation including medicalised death has informed both medical anthropology (the discipline which underpinned the study reported here) and the sociology of health and illness for several decades (Broom and Woodward, 1996). Stacey, amongst others, objected to how patients (and by inference, their bodies) came to be seen as the 'work object' of medicine (Armstrong, 2002). Douglas (1966) considered that in most cultures the body is used as a social symbol reflecting the technologies of the society in which it lives – often a machine separated from the self, a machine that can ideally be fixed which is how many deaths have come to be framed and scripted in contemporary society (Katz, 2001: 7). Indeed, Clark (2002) re-visits Illich's 'medicalisation of dying' critique (1976) and suggests only limited progress has been made, despite the rise of the palliative care movement. He states that medical services (and society more generally) continue to regard death as something to be resisted, postponed or avoided. CPR would appear to be one way in which this is currently being achieved and the detail of this is explored next.

Cardiopulmonary resuscitation

Generally, CPR of an adult may last between 5 and 15 minutes; although depending on the particular circumstances the time period may be much longer. The goal is to save a life, or, more realistically perhaps, to delay death. This may involve securing the airway of the 'patient'[6] often via intubation techniques, delivering oxygen, monitoring of the cardiac rhythm, applying chest compressions, securing intravenous access, giving drugs and passing an electric current through the heart. The event, no matter how smoothly managed by attendant staff (which is often not the case), is at best invasive and dramatic[7] and often does little to change the ultimate reality borne out by the survival statistics (Cooper *et al.*, 2006). As Sokol (2009) notes, many non-clinicians hold rosy views about the nature and effectiveness of CPR. He suggests that several studies have underlined their misplaced optimism: in one, the 269 respondents reported a mean expected survival rate of 65 per cent; in another, 81 per cent of respondents over 70 years old believed the likelihood of leaving the hospital after a cardiac arrest to be at least 50 per cent. The real survive-to-discharge figure for all in-hospital cardiac arrests is roughly 14 per cent, and many survivors will have functional or neurological impairments, with considerable negative impact on their quality of life. For those who have suffered an out-of-hospital arrest, the 'success rates' are even less encouraging; and if initial resuscitation has been delayed in any way, the prognosis is particularly poor (Brummell, 1998).

Timmermans' work, a sociologically constructed account of CPR events observed within two American A&E departments, highlights the success of the resuscitation movement in enrolling entire societies in the belief that every sudden death is premature, '...conditioning people to massage chests and put their mouths on the mouths of strangers was one of western medicine's major achievements in the late twentieth century' (1999: 30). The accomplishment of the resuscitation project was to turn life-saving into a taken-for-granted activity, to make resuscitation a morally sanctioned routine in which one and all are caught up legally and morally in an obligation to intervene effectively. He shows how socio-political influences have and continue to reinforce the ethos of 'Saving Lives'. This has promoted resuscitation to a position whereby it has grown out of all proportion to its original intentions and how, despite the rhetoric, its universality as a procedure ('anyone, anywhere'[8]) is questionable given the appalling outcome statistics.

Despite these statistics the resuscitated death appears to be on the increase and is no longer confined to either the hospital emergency department or other specialised health care arenas. Indeed the tentacles of the resuscitation movement extend throughout the hospital and far beyond its walls and have come to affect a particular view of how death should be managed. Principally, this is by preventing or delaying it (McNamara, 2001: 69). The significance of this movement and the cultural acceptance of resuscitation as a measure to avert death are evidenced in the community-wide resuscitation training programmes, public access to defibrillators (to re-start the heart) and the advent in every hospital of a specialised group of health care workers, Resuscitation Training Officers (RTOs). This occupational group are testimony to the professionalising and medicalising of death commented upon so much in the literature,[9] their role being (uniquely amongst health workers), the preparation of people and of equipment for one intervention: CPR.

Methods and data

I 'shadowed' this group of professional RTOs, in two acute hospitals in the United Kingdom for the 3-year duration of the study, in the different areas of the hospital as they undertook their daily checks on the resuscitation equipment and observed them teaching other staff how to resuscitate. Access to the CPR event itself (in one hospital) was eventually agreed on the understanding, (by virtue of my clinical background), that if necessary I would help at the emergency situations I was observing, that is to say, become more of a participant than an observer.[10] The field notes below reflect just such a situation, but they are included here as a means to demonstrate that the hospital is a place of dying, the type of space within the hospital where deaths may occur (A&E departments) and the material artefacts common to such spaces – material objects that come to represent or form (quite literally with CPR) extensions of the body (see Hallam and Hockey, 2001).

Accident and Emergency department of a District General Hospital in the Home Counties, England: Field notes from an observation day

> *[The] arrest bleep went off and alerted us to a patient soon to arrive in the Accident and Emergency department [an arrest = an individual whose heart has stopped beating and is undergoing resuscitation procedures]. We went off down the stairs to A&E. The ambulance had yet to arrive. KR [the RTO I was 'shadowing' that day] said something to the staff about who was taking this one? Various members of staff were milling around in the resuscitation bay. I'm not sure who everybody was but in the end a female Dr asked if she could take the arrest when it came in. KR asked if I wanted to go and meet the ambulance. I felt very nervous – so divided in my roles. I didn't feel like a senior nurse used to taking command of clinical emergencies (the old me). However, knowing that my access and my research credibility were all predicated on the competent ex-ITU nurse identity, I agreed to go with him. I copied KR by putting on a plastic apron and latex gloves and went to stand outside the doors of A&E to wait for the ambulance to arrive. It is a highly visible thing to do and I was aware of people in the waiting area watching us and thought how awful for them to be able to see the incoming patient and, at the same time, how exciting. I recognised a feeling from long ago of privilege at being able to be part of the clinical world, perhaps even a misplaced sense of self-importance at being involved in a medical drama. It is difficult not to feel like some sort of hero figure when you are waiting to see if you can save someone's life.*
>
> *The ambulance arrived and the paramedic got out of the back and spoke to KR. I saw the patient, an elderly man of 83 years of age who very clearly looked dead to me. He had apparently had 9 electric shocks in the back of the ambulance (highly unusual, KR later commented that he hadn't seen 9 shocks given for a long time) and 5 shots of adrenaline and still no response. He had been intubated and cannulated en route and received a dose of lignocaine to try and treat his chaotic heart rhythm (ventricular fibrillation (VF) – the current drug of choice, Amioderone not being available). From his VF he then went into a period of asytole (no electrical activity). I went to help bring the stretcher in and noticed KR making a somewhat half hearted attempt at cardiac massage. I felt very uncomfortable that I hadn't even thought to do it – the chap was so obviously dead that it never occurred to me. His eyes were open and he had apparently 'been down' for 45 minutes.*

The paramedic said that the wife was sitting in the front of the ambulance and that he (or his colleague) would bring her in. It was her that had found him apparently. The man was brought through into the resus area and ECG leads put on. The rhythm looked like fine VF or was it artefact? Several Drs and 2 anaesthetists (at least) were there by then and suggesting the 'gain' on the monitor should be increased. I think KR attended to this. I was aware that the arrest wasn't running very smoothly, no one seemed that interested or assumed a prominent leadership position. This made me feel uncomfortable. People were milling around not actually getting on with things. I recall talk of the algorithm cycle – 3 more shots of adrenaline and 1 of atropine was given I think. One senior anaesthetist who looked very young and cool sat in the background watching, making comments about what should be going on. JK, the A&E consultant, came in and in his rather bombastic manner starting asking questions and generally increasing the tensions in the room. KR asked if I wanted to do cardiac massage, so I did. It felt absolutely awful. I have done it many times but never felt anything like this. I felt I was violating this man, this body in some way. He was so very obviously dead but more horrible than this, when I put my hands on his chest, his chest simply caved in. There was absolutely no resistance at all. It felt quite ghastly and was the result of all the electric shocks and cardiac massage his elderly, frail chest had already received. I was being ineffective, I could tell by the cardiac monitor that I wasn't producing a cardiac output and one of the anaesthetists asked if there was a stool for me to stand on (he had also noted my ineffectiveness and wanted me to be in a position where I could put greater pressure on the chest). I felt embarrassed that I wasn't doing it properly but was so horrified by the state of this chap I simply couldn't find the adrenaline surge I needed to. In addition I was soon aware that I was massaging slightly too slowly and not continuously – he was intubated so I didn't need to leave a break for lung expansion. I felt silly that I had forgotten all this. The shocks had clearly 'frazzled' him and I think his sternum was broken (KR agreed) so it was like putting your hand through someone's body. I asked KR to come and take over which he did. Someone asked 'where are we?' meaning where are we within the prescribed protocol/3 minute cycle. I was certainly unclear. I was aware that one of the senior nurses, a male, heavily tattooed, an ex marine apparently, was making light hearted conversation about all going out for a drink on the 'ash cash'[11] *that would be coming from this arrest. I think he asked at that time which of the doctors was going to come and speak to the wife. There was general agreement after a short while that this chap was unsalvageable. Everybody disappeared very quickly and I noticed, once again, that one nurse was left to clear up the body and the mess in the clinical area before his wife was to be brought in. As I walked out into the main A&E area I heard the nurse ask if she was taking his wedding ring off or leaving it on. At the nurses' station, the male nurse (tattoo) had previously been talking to the wife and ascertained that the chap was a Roman Catholic and was telephoning for a Catholic priest to come down to the resus bay. KR looked at the ambulance records and started to complete an audit form noting down times of call, time to first shock etc. etc. I listened as the male nurse then called the patient's son to come to the hospital. He said the situation with his father was very grave but stopped short of saying he was dead. I wondered about this for ages and wished I had had the opportunity to ask if this was policy. I also remember speaking to a radiographer in A&E at the time. A middle aged, well spoken woman who, when discussing the resus, said how pleased she was that her own father had died at home and not had to undergo any of this.*

The field notes above relate to a typical medicalised death occurring in a typical impersonalised medical space (A&E), yet in medical terms, comparatively high status,[12] which is set out and equipped specifically to facilitate the practice of medical emergency work. It is a place replete with 'body-related' medical artefacts, that set the tone for the important work of life-saving that takes place here, artefacts that remain largely constant juxtaposed with a myriad of non-constant bodies (e.g. staff, patients, ambulance and security personnel, police, visitors, relatives and students) who come and go, dead or alive, whilst the boundary of life and death is routinely transacted and mediated all around them.

The resuscitation bay of an A&E department creates a particular social space for framing how death and dying are constructed and the efforts that modern medicine makes to try and avert it. An acute hospital ward, especially one designated a coronary care unit as in the subsequent set of field notes, does the same. Both resuscitation bay and hospital ward are public spaces but populated by cultural objects (in their own way reassuring and necessary) which denote the key business of the place – medical work – primarily to heal people who are sick and avoid death. Cardiac arrest trolleys, ECG machines and portable defibrillators are clearly visible at all times, each redolent with associations of clinical intervention, power, the primacy of life and the bio-medical frame of it all. The material culture in the unit (cardiac monitoring and so forth), as with the resuscitation bay of an A&E department, both reflects and shapes the practice that occurs within that space, designed as it is, to help control and police the liminal space between life and death. Whilst Hallam and Hockey (2001: 203) suggest 'Death calls for the production and use of dedicated materials', the work of avoiding death equally calls for as many.

By the time I arrived at the CPR event detailed below, the patients' bedside curtains were semi-drawn and the normally quiet, uncrowded bedside space had become a crowded emergency medical treatment zone, semi-shielded by the (flimsy and ill-fitting) drapes from the curious or frightened eyes of other patients. As with the event in A&E, those who congregated in the space around the dying body were professional staff not relatives or people close to him. They assumed the role of informed audience although a number were also called on to act. In this case I was only audience.[13] I could see the body throughout the action but was not called upon to act. Other patients on the unit formed a quasi-audience, acutely aware of the drama unfolding in their midst but had to rely mainly on aural and the occasional visual clue to understand how the action was unfolding.

I immediately noted that the victim was still struggling for life. This was evidenced by movement of the limbs and the gasping, choking sounds being emitted from the airway. Eleven staff were already in attendance. The space around the bed was crowded with people and an array of equipment and monitors. The clinical picture appeared complex and the staff were puzzled by this interesting 'case'. The resuscitation procedures already underway were being conducted in a matter-of-fact manner due in no small measure to the calm and knowledgeable actions of the experienced nursing staff present. However, the recalcitrance of the body in question and its refusal to be 'mended' generated a degree of tension manifest in the clipped tones of the senior medic directing procedures, particularly noticeable in his overriding of suggestions offered by other team members. The following two paragraphs are an amalgamation of lengthier field notes from that day.

Field notes

When we got to the arrest I saw the patient lying on his bed in his pyjamas although these were pulled down past his groin. (I later noticed he had been incontinent of urine). He had one slipper on and the other half off. He was pale but breathing and making some movements. Two women in theatre greens were up at his head. In addition, there was one female and two male doctors standing by. The female doctor spoke a little, asked a few questions. She and another male doctor sought guidance from the male leader; the other doctor present. A male staff nurse was doing chest compressions. A female student nurse was standing to one side, quietly and calmly helping with the ECG monitoring and such like. There was a senior female staff nurse giving drugs to the patient in accordance with directions from the senior medic. A male staff nurse was moving in and out and helping on the periphery. I noticed a staff nurse I knew on the unit. She recognised me and came and spoke which I found very reassuring. MSt. was also present dealing with equipment. MA watched the arrest carefully – it was quite extraordinary to see the patient slipping away in front of our eyes.... He was defibrillated 17 times which was quite awful to see. [in terms of the 'assault' on the person this represented, the somewhat desperate nature of the resuscitation and the 'disrespect' to my mind of death and dying].

I could feel the man dying in front of me and had a sense of him moving on. He was surrounded by 13 staff (few, if any of whom knew him well) plus four other patients were very close by, none of course were family.

The action lasted approximately 20 minutes. When all efforts failed the attending medics disengaged and disappeared in a matter of seconds. The body was left cold, pale, bare, exposed, eyes open and covered in bits of litter from the resuscitation. The staff member I was with closed the man's eyes and asked for a blanket to cover him up. There was an odd temporal space left by the departing medics prior to the nurses returning to prepare the body for viewing and then transfer from the ward. In this space nothing much happened, which was a stark contrast to the frenzied activity of only moments before. The disappearance of the staff (except a male nurse) seemed to indicate that the important things were over, what was left was unfortunate, of little interest, matter out of place in an acute clinical environment, unwelcome. The actual death, the lifeless body, seemed, although not an entirely unexpected outcome, nevertheless, an unwanted end to the CPR performance just enacted. The departure of the medical staff also emphasised that the performance (about to commence) of preparing the corpse and managing the body was certainly not their work. I thought of the things I had read on death and pollution and how this was considered low status work (Lawler, 1991).

One medic remained on the unit, away from the body, to write up the notes. He discussed with me and MA the clinical condition of the patient who clearly had been a very sick man. His comments were interesting; a sort of rationalising process where the death was now being constructed as unavoidable (not our fault) given the severity of the clinical circumstances.

Later in the day, MA and I revisited the unit. The atmosphere appeared subdued not helped perhaps (according to one nurse) by another patient complaining vociferously about the body still being on the ward. I asked this nurse what he had made of that mornings event. He commented on what he perceived to be the sub-optimal airway management, the potentially dangerous handling of the defibrillator equipment by the cardiac registrar and leadership problems. The identity of the person concerned, the patient, associated family connections, any effects on the staff and the tragic outcome, were largely absent from the discussion. The focus was the performance of staff.

Space and place

My observations suggest that the resuscitation bay of an A&E department and an acute hospital ward (especially one designated a coronary care unit) create a particular social space for framing how death and dying are constructed and the efforts that modern medicine makes to try and avert it. Throughout the events the space is one of contained, time-limited, intensive action with the centre-stage yet 'out of control' body linked directly to the surrounding artefacts and, by virtue of their actions, the actors within that space. It is a space in which a life is saved or lost, medicine and technology win or lose and death, as an outcome, is somehow made easier by virtue of the clinical environment and the drama (CPR) that is enacted around it.[14]

O'Toole and Were (2008) comment on the dearth of attention paid to the physical layout, the material culture/objects and the sense of 'place' that form such an important part of the context of all qualitative research. They draw on Schlereth (1982) to emphasise that it is through material culture that we can learn the 'belief systems – the values, ideas, attitudes and assumptions – of a particular community'. For example, the material culture associated with the high-tech, life-saving world of in-hospital CPR, the places, spaces, actors and equipment so often depicted on hospital TV docu-drama or 'soaps', testifies to the power and position of bio-medicine in western societies and the belief systems that underpin its authority. As O'Toole and Were (2008) suggest, objects have meanings based on culture, function and power and their role is to 'provide a framework where action takes place'. This can be seen in the 'medicalised death' contexts described here where the interaction between the material culture (objects such as defibrillators), the involved participants (the medical team, the body that is being worked on) and the place in which this is being conducted (hospital A&E department or coronary care unit) are intrinsically interlinked and reflect issues of power, identity and status. It is also where the dead body, transformed from the container of life to the container of death, becomes medically invisible for although death is a relatively common event in these public spaces, it is unwelcome. Paradoxically, this is where most deaths (in western countries, at least), occur. Indeed it is in hospital that the boundary between living and dying is constantly challenged, and the space where that often occurs is in A&E. In the coronary care unit (like that cited here) patients who are considered stable are placed on the 'ambulant/self-caring' wing of the coronary care unit prior to discharge. Death, although not entirely unexpected, is relatively unusual here. The focus is on rehabilitation and living (see Chapter 4). The second death described was constructed (by the medical staff) as unfortunate, untimely and wrong in some way and the corpse itself was clearly considered by other patients to be occupying the wrong space. CPR had created a liminal space between life and death, but with death as the outcome, the corpse was very much 'matter out of place' (Douglas, 1966), an ambiguous '... presence that manifests an absence' (Vincent, 1991, cited in Hallam and Hockey, 2001: 14), an absence of a life, an absence of medical success.

Death (once medically sanctioned) changes the nature of the space; the action stops, the medical team move away, attention is diverted elsewhere towards the lives of other patients. The space including the body within it is cleaned and re-ordered; equipment and detritus are removed and as soon as time allows, the body is prepared for relocating to the mortuary. The well-documented ritual associated with removing the dead body from the ward unfolds, and the now-empty space is prepared for the next patient, the next body, the next occupant. Thus a series of transitions occurs; a patient whose life is being saved moves to organic waste that evidences medical failure and needs cleaning away, to a body prepared for viewing and then (on the ward) to re-usable (sanitised) bed space for the living. It is the re-usable space, that brings in, albeit also transiently, its own emotion, embarrassment; itself a manifestation of the materiality of absence.

Conclusion

The cultural meaning associated with both types of hospital space highlighted here is that of life-saving by virtue of medical intervention. Much of the material culture within these spaces is associated with the boundary between life and death, death-related objects (as well as the body itself) which are part of the fabric of hospital life, used by those skilled in their manipulation to try and order (potentially) dying bodies. The emphasis throughout is on life, life-saving.

The original research from which this work is drawn sought to explore CPR as a (potentially problematic) cultural treatment of dead and dying persons. In this chapter the relationships between the spaces, places, objects, bodies and actions associated with CPR-mediated deaths have been singled out for particular attention. What is more, the rupture that CPR represents both to the victim, their friends and family and to the flow of hospital life serves to expose both the power and the limits of medicine and the overlapping and sometimes-conflicting agendas it, and the hospital space in which it does its work, face: life and death.

As Turner reminds us, 'Disorders related to stress and lifestyle do not lend themselves to simple solutions through what we might call 'heroic medicine' (1995: 215). Faced with the challenge this poses, it is possible that CPR presents an opportunity (albeit unconsciously) to reaffirm the power of medicine and its relationship to ideas about the body, life and death. As such it reflects deep cultural themes. It appears to be more than simply a useful life-saving procedure or a pertinent focus for an ethical debate, and the rapid expansion of CPR techniques certainly cannot be explained in terms of its clinical or cost-effectiveness. In an era where the doctor is said to have replaced the priest, CPR seems to hold a marginal yet significant and symbolic role in health care practice that is over and above the way it is usually explained in the literature or discussed in everyday practice. Despite its statistically overwhelming association with death, CPR retains its status as a life-saving or indeed life-giving technology. As such, the dead bodies it so often produces have no place within its spheres of expertise and practice. Hence the symbolic invisibility of the corpse in the eyes of those administering CPR, a focus only for the attention of lower grade nursing staff.

CPR is perhaps an unsurprising outcome of bio-medical dominance; its protocols and procedures provide guidance for a meaningful death and bestow on the death a particular set of understandings, that is, everything possible has been done to avert it. The performance, be it in the preparation for managing death (CPR skills training) or in the actual event itself, serves as a marker by drawing attention to the significance of what is occurring and in actually undertaking the actions, attention is diverted away from the proximity and finality of death. The performance itself appeases quasi-legal concerns by showing committed intervention, it offers a degree of solace to the family that everything was tried and it reinforces for the staff their role in these circumstances.[15] Interestingly of course, this role (of CPR provider) can be practised and re-practised, adjusted and perfected, whereas the outcomes cannot.

Therefore, the notion of performance appears to underpin much of what is happening in CPR, and the material above highlights the double meaning of performance in this setting – performing as accomplishing, and performing as in acting within a ritual (which has mainly symbolic rather than functional efficacy). It is a performance which, more often than not, provides a symbolic marker of the passage from life to death, a passage mediated by professionals within a bio-medical frame of reference.

Part of the medical frame of reference is derived from attempts to survive disease and surmount death (Romanucci-Ross *et al.*, 1997: ix). Thus, CPR may be viewed (on one level) as a modern-day solution to the 'problem' of sudden death in an acute health care environment. Predicated on a belief in the body as a machine that can be mended, that sudden death or catastrophic clinical events are reversible, it appears to demonstrate (particularly given what we know about the outcomes of CPR) a privileging of concern with the physical body over the social body. However, this is likely to be only part of the story. An alternative perspective might perceive CPR in terms of a symbolic ritual which provides a particular way of thinking about death. As such it is illuminative of the culture in which it occurs. A re-framing of CPR practices in this way might also highlight the complexity of the relationship between expectations of health care and health care delivery, for whilst CPR is clearly a biological / medical event it is also a cultural one.

The potent, death-denying, life-saving image and activities associated with resuscitation may appease a cultural disquiet about mortality, for as Bauman (1992) argues, one of the life strategies which characterise contemporary society is the deconstruction of mortality into battles against specific diseases or health risks. On a more pragmatic level, the act of (at least an attempted) resuscitation has come to be offered as a defence against an ever-more pervasive legal system, widely perceived, rightly or wrongly, amongst health care workers to be ready to police medicine's duty of care and professional accountability. Thus, for those who are neither inclined to voluntary euthanasia nor being cared for under the umbrella of palliative care, the resuscitated death and the potential horrors that may accompany it are increasingly impossible to avoid. That is to say, the 'professionally devised plot' (Del Vecchio Good, 1995) is difficult to get out of since 'all humanity is a possible candidate for resuscitation' and its current position is now 'part of the final, common pathway to death' (van Delden, 2000: 240, 249). How this has come to be and how this is enacted and sustained underpin the ethnographic work reported here.

Notes

1. See National (NHS) End of Life Care programme: www.endoflifecareforadults.nhs.uk.
2. A cardiac arrest and subsequent CPR can occur as a result of a number of different clinical situations, for example, trauma, hypoxia (i.e. lack of oxygen as a result of, say, an asthma attack, drowning, choking etc.), sepsis, hypovolaemia (i.e. low circulating blood volume following dehydration or haemorrhage), electric shock, drug overdose, electrolyte ('chemical') imbalance and notably in the older adult population, from cardiac arrhythmias such as ventricular fibrillation, heart attacks (myocardial infarction), ischaemic heart disease and cardiac or multi-system failure.
3. An example perhaps of what Fox (1997) terms 'one-to-one congruity between architecture and the function of a space'. Resuscitation bays certainly are spaces ascribed with particular meanings and facilitate certain behaviours and routines.
4. A phrase attributed to the cardiac surgeon Claude S. Beck who convinced the world of the merits of electric defibrillation to treat the life-threatening heart arrhythmia ventricular defibrillation. See Timmermans (2002).
5. This refers to equipment used to make sure those who have suffered a cardiac arrest or similar event are able to receive adequate oxygen/ventilation.
6. The use of the medical term 'patient' is deliberate here since it reflects the context in which CPR was undertaken and that in which the ethnography was conducted.
7. Davey (2001) refers to it as a 'violent' intervention.
8. A well-known and oft-cited phrase linked to CPR practice drawn from a landmark article on CPR (Kouwenhaven and Knickerbocker, 1960).
9. Unlike, say, bereavement counsellors, Resuscitation Training Officers would not describe themselves as death workers; despite their close association with death their role is constructed around notions of life-saving.
10. My observation data were supplemented in the usual manner by field notes, a research diary, a number of 'formal' (audio-taped) interviews, several unrecorded interviews (as requested by participants) and many 'on-the-job' conversations.
11. 'Ash cash' is a universally recognised term (medical slang) referring to the morbid practice of receiving payment from filling out cremation forms.
12. As indicated, the 'resus' [citation] bay is where the most critically ill patients are taken and where high-technology life-saving interventions are routinely used.
13. It is appreciated that my shadowing of the RTOs wherever they went within the hospital, sometimes observing, sometimes participating, was also a manifestation of occupying a particular space.
14. Interestingly, both Seymour (1998) and Timmermans (1999) question the frequent placing of technology in the role of 'bad guy' around death given its use is now so commonplace. Moreover, its negative profile is increasingly incorporated into contemporary constructions of a natural, dignified or meaningful death (Bradbury, 1996). However the work of Lock (2002) draws attention to the on-going disquiet and myriad of still unanswered questions the role of technology has bought to the bedside. Nonetheless, Mohammed

and Peter (2009) see CPR (alongside other medical interventions) as a 'moral practice' which offers a ritualistic benefit to patients, families and health care providers, helping to facilitate the process of dying; its social value being as important, possibly, in the medical futility debate as medical scientific criteria.

15. Seymour's UK study (1998) also identifies the role of 'we did everything possible' as a significant explanatory position adopted by intensive care nursing staff when dealing with the death of a patient.

3
Making Hospice Space

Ken Worpole

This chapter explores the spatial aspects of the contemporary hospice, particularly in the way its architecture and landscaping are structured to shape and manage the process of dying – for the patients themselves, for close family and friends as well as for staff. It is based on the author's study of more than 20 hospices and cancer care centres in the United Kingdom, Ireland and Scandinavia in 2007 and 2008, while researching the book, *Modern Hospice Design* (2009). It draws conceptually on twentieth-century writings on 'social architecture', particularly the development of new building types associated with health, social welfare and the rise of the modern welfare state. However, it also draws on contemporary understandings of the nature of 'sacred space' – and certain religious traditions informing architecture and landscape – which are brought to bear upon this particular 'end of life' building type, now becoming an increasingly common element in modern townscape and social provision.

The place of death

Sooner or later changing social concerns find permanent form in new building types, one of the abiding themes of architectural history. In the early part of the twentieth century, the rise of more democratic and collectivist politics brought with it a plethora of pioneering buildings and public settings, ranging from welfare clinics, nursery schools, neighbourhood libraries, community centres, recreation grounds, open-air lidos and other facilities associated with programmes for improved public health and physical well-being – particularly for children. The term 'social architecture' has been coined to describe this particular range of building types (Saint, 1987; Worpole, 2000). These have made

Figure 3.1 This 1936 plaque unveiled on 25 September 1937 at the opening of the new Public Health Service Department in Southwark, London, exemplifies the municipal commitment to public health.

a permanent contribution to the form and appearance of the modern town or city, and provide a continuing reminder of changing social expectations and collective endeavour. Such built forms also physically embody and 'remember' a politics associated with communal welfare – and even a celebration of life – expressed through the social construct of the healthy body and its centrality as an achievement of public social policy (Figure 3.1).

Rather less has happened with regard to the other end of life: that is to say, in providing places and spaces of end-of-life care for ageing and dying people. While the sanatorium or the open-air school became iconic buildings of modern movement architecture, there was no such innovation in the design of nursing homes or places for the care of older or terminally ill people. Such concerns may have been regarded as belonging to outdated and morbid forms of religiosity, or a pre-occupation with the eschatological, and therefore beyond the remit of collectivist welfare policies. Even today few politicians or architects wish to talk about places of dying within the wider community, or find any compelling requirement to put it at the centre of public policy, as has been evident in Britain in recent decades with the failure of governments to act over the continuing decline of cemeteries, and the need for political debate and action on allowing the re-use of graves (The Stationery Office, 2001).

One notable exception to this political and architectural indifference was the rise and subsequent ubiquity of the municipal crematorium across Europe (though not North America) in the twentieth century. This was part and parcel of a related rationalism around the 'democratic body' in public policy and provision, in which town planners were among the most enthusiastic supporters of cremation for reasons of land use, fearful that otherwise the major cities would eventually be surrounded by a 'white belt' of cemetery land, rather than a green belt of natural amenity providing public access to rural landscapes. Even today, these 'unhomely' places are still regarded in some quarters with a degree of distaste or social embarrassment, and remain relatively marginal to the architectural and social history of modernity. At least this was the case until the publication of Grainger's (2005) fine architectural history and gazetteer of this extraordinary new building type.

Nevertheless, the crematorium has now earned its rightful place in the twentieth-century canon of social architecture, and to this impressive portfolio of new building types must now be added something equally different and socially new: the modern hospice. It is typologically and culturally a new phenomenon which, nevertheless, has roots in long-standing pastoral and religious traditions associated with Christian, Jewish and Islamic beliefs and practices. Its rise indeed might be part of the re-discovery of the 'spiritual' in matters of public welfare, in what is now often described as 'post-secular' (or post-rationalist?) society. The turn towards an acceptance of complementary medicine within public health provision is another aspect of this shift.

The modern hospice movement is commonly agreed to have come into being with the opening of St Christopher's Hospice in Sydenham, London, in 1967 (Figure 3.2).

Figure 3.2 Entrance to St Christopher's Hospice, Sydenham, opened in 1967.

Its founder, a devout Christian, Cicely Saunders (later Dame Cicely Saunders), was a nurse who was subsequently trained as a doctor. She resolved to do something when a patient she had befriended in 1948, David Tasma, a Polish refugee, discussed with her the idea of establishing more homely places where people could die in greater tranquillity than was currently the case. When he died in February 1948, he left Saunders a bequest of £500 in the hope that she might establish such a home, with the now famous appeal: 'let me be a window in your home.'[1] That window now exists in the entrance foyer of St Christopher's.

From the outset, the hospice ideal was regarded as requiring a new and inspiring architectural form: 'Saunders and Tasma imagined a *house* where people could find relief from pain, where they could meet with encouragement for self-awareness and socialisation, and where the setting would be uplifting, not depressing' (Verderber and Refuerzo, 2006: 15).[2] This attachment to the domestic in both the scale and the form of the hospice building has, as I argue, become a generic feature of these new spaces and places for dying people, which are now also appearing in many other parts of the world. For example, it can be seen in the first self-designated hospice in North America, Connecticut Hospice in New Haven, opened in 1974, or the first hospice in Denmark, Sankt Lukas Hospice in Hellerup, which opened as recently as 1992.

Since its inception by Saunders, the movement has spread widely. There are now over 250 dedicated hospices in the United Kingdom, and 8500 modern hospice projects in 123 countries across the world, and the number is growing all the time (Worpole, 2009). While this proliferation has been rapid and remarkable, it has not been without resistance. In Denmark, for example, there was an initial political and indeed public reluctance to entertain the idea that there might be special places created by the state where people would choose to go to die. Indeed it was said when the idea was first mooted that ' "hospitals of death" do not fit into the Danish mentality.'[3] That has now changed, as is evident from the recent publication of the first fully developed national set of design principles for hospices in Denmark.[4] Furthermore, in December 2004, the Danish Parliament passed legislation obliging all counties to develop plans for a minimum of 12 hospice beds in that county.

The idea that the modern state might create places where people could go to die still remains potentially disturbing to many – illustrated by the comments recorded in Denmark. In the United Kingdom, and in most other places where the hospice movement is expanding, the prime agencies remain religious organisations, usually allied to or working through the organisational structure of local charities. In 2006 it was stated that though the hospice movement was first developed in the United Kingdom by voluntary hospices, who continue to provide the majority of in-patient care, there is now a closer partnership with the NHS who manage about 25 per cent of in-patient palliative care services and provide funding for about 34% of the cost of voluntary adult hospices (4 per cent of children's hospices).[5] In future we might expect to see closer working between public and charitable bodies in providing for people who are dying. However, it is still a matter that is fraught with personal anxiety as well as political discomfort. This is particularly so when discussions about the management of death in hospices are increasingly, though erroneously, elided with more interventionist practices such as 'assisted dying'. It is unlikely that such discomforts or procedural ambivalences will be resolved in the near future.

Thus there remains public apprehension that death could be so tidily managed. The matter of where people die and how they die is a subject that is still broached with some difficulty. It is evident that the place of death is understandably imbued with great symbolic import and deep personal meaning, particularly among bereaved people. This is creating new problems in that whereas people used to die at home, the majority now die in an institution. At the beginning of the twentieth century in Britain, it has been claimed, 'about 85 per cent of people died in their homes, with workhouses accounting for most other deaths. By the mid twentieth-century around 50 per cent of people died at home. In the early twenty-first century acute hospitals have become the most common place of death.'[6] In Ireland, around 67 per cent of deaths occur in acute and general hospitals, and in North America the figure is now 80 per cent (Worpole, 2009).

Of the 58 per cent of the UK population who currently die in hospitals, such deaths mostly occur in large public wards, where there may be little professional expertise in end-of-life care (Worpole, 2008: 9). At present only 4 per cent of people in the United Kingdom die in hospices, though many more would if they could, we are told, and availability is still regionally biased: hospice deaths vary between 2.5 per cent on the North West of England and 5.9 per cent in the South East, providing some indication of availability of provision (Department of Health, 2008: 25). However, even what is classified as 'home' for people who are dying is debatable, since for those 20 per cent of the elderly population resident in care homes, the room or ward they now inhabit is *de facto* their home (technically their permanent residential address). Most people, however, would still regard nursing homes as institutions (see Chapter 5).

While dying at home may be an expressed ideal for many, these attitudes can quickly change once serious pain or familial anxiety sets in – or, more positively, when people experience good hospice care on a temporary or respite basis and consequently change their mind. Payne (2006) claims that 'Dying at home is very scary. You might think you want to do it, and you might want to support someone in your family who says they want it, but in the midst of it actually happening you may feel you'd be better off elsewhere.'

Such is the growing political concern over the conditions in which many older people die, whether at home or in a hospital, that an All Party Parliamentary Group on Dying Well (2009) has been established in the United Kingdom. The very idea that parliament should discuss the concept of 'dying well' suggests that we are entering a new era, in which death – and the place of death – returns to the centre of social policy. Where social policy leads, architects and builders shortly follow, and it is the architectural and cultural space of the hospice as a new and privileged setting in contemporary society which is the principal matter of this chapter.

To have addressed the public fear of death in an institution in such a short time is no small achievement, and this is what the modern hospice has accomplished. For it is still the case that for many older people the experience of hospitals – especially as places where one might spend one's last days – still carries a long history of popular memories and mythologies surrounding the typically high-walled, blackened brick Victorian building on the hill (in England, often the former workhouse), sometimes adjacent to the cemetery, though now embellished with a new corporate notice-board: grim, bureaucratic, prison-like in its design and ethos. When I was involved in an oral history project in Hackney in the 1970s, I interviewed many older people who expressed a great fear of ending up in Hackney Hospital as it then was which had been the former borough workhouse (Figure 3.3).

That the treatment of terminally ill people is still dealt with unsatisfactorily in hospitals is revealed by the fact that over 50 per cent of all complaints in the UK National Health Service relate to the care of a patient who is dying. It would appear that hospitals are still not good places in which to die. Concerns over dying in hospital were reinforced further by a report published by the UK Healthcare Commission on 27 September 2007 based on visits to 23 hospitals. This found that only five complied with all the government's core standards for dignity in care.[7] When it is considered that in England people aged 65 and over constitute 16 per cent of the population but occupy nearly two-thirds of

Figure 3.3 St Leonard's Offices for the Relief of the Poor, the former Shoreditch workhouse and hospital in Hackney, built in 1863. Now an out-patient clinic called St Leonard's Hospital.

general and acute hospital beds, the scale of the problem becomes even more pronounced.[8] A later survey published in April 2008 claimed that 'Fewer than half of the staff in NHS trusts across England think caring for patients is the top priority at their place of work.'[9] It is apparent that as a result of the sheer numbers of older people now dying in hospitals, close personal care at this crucial *rite de passage* is found wanting and inadequate to the occasion.

Such conditions seem to support the view of French architect, Degrémont, who has provocatively suggested that 'In our civilisation, a death outside the home can be likened to a death outside society – because, in a hospital environment and despite the existence of palliative care units, death is still something of which to be ashamed' (1998: 127–129). What the architecture of the modern hospice seeks to create at the outset is an appearance and atmosphere that is as far from that of an institution as it is possible to be. To achieve this it is evident that other institutional models – religious retreats, settlements, care homes, hotels even – have been considered and learned from as potentially providing a template for a building type that is fundamentally new, even though it now appears so well established.

Suburban borderlands

One of the problems in articulating the main spatial elements of the modern hospice – as a place where death occurs in a relatively ordered and managed way – is that these settings are subject to continual adaptation and change. The early hospices such as St Christopher's were principally designed to provide accommodation in small wards or single rooms for a restricted number of patients in the final few days of life. This setting was clearly understood by all concerned – patients, family, staff and visitors – to provide no outcome other than the imminent death of the patient in as dignified a manner as possible. In architectural form they emulated the cottage hospital, with the addition of a dedicated chapel and an ornamental garden. The processional aspects of this were relatively straightforward: patients went in at the front door and left at the rear, via the viewing room and morgue. This is now changing.

Today the provision of bedded accommodation is becoming less of a core feature of hospice buildings or settlements. Increasing numbers of patients are attending on a day-patient basis, coming for clinical consultation, therapy, pastoral care and emotional support; residential patients are now in a minority. The average modern hospice may provide beds for 20 patients but may be treating more than 400 out-patients, mostly in their homes. Princess Alice Hospice in Esher, for example, contains 28 beds, but cares for 2,000 people each year. The one-way door has become a revolving door, and to some extent the hospice which was only a short while ago configured as an intimate and quietist antechamber to death has become more like a village fete where people come and go, even though many are conscious that they are in the last weeks or months of their lives. The opening of the New Farleigh Hospice in Chelmsford in 2006 was significant in that it was the first new UK hospice to be designed and built without a mortuary, as more and more patients return home to die, and much of the focus of the hospice is on the care of day-patients.

Nevertheless, there are a number of significant architectural and spatial elements which make up the basic form of the modern hospice setting, and these will be dealt with somewhat schematically here. An obvious starting place is the location and public face of the building within the wider townscape or landscape. For Cicely Saunders the ideal location for a hospice was within a suburban setting, becoming a 'good neighbour' to local residents, but taking advantage of public transport connections and providing easy local accessibility. From the outset, hospices were envisaged as community buildings drawing their patients from a defined geographical catchment area rather than being planned as regional or specialist facilities open to anybody terminally ill no matter where they lived. The suburban setting reinforced the motif of the domestic/communitarian style of building, providing a borderland space which was neither part of the bustling, noisy city, nor regarded as a rural retreat or remote fastness. Such small, residential units fitted well within the 'dormitory' ethos said to characterise aspects of suburban life. This suburban 'borderland' setting seemed appropriate to the 'borderland' ethos of the hospice, which patrolled and controlled the lonely land situated between life and death, as well as night and day.

Since that early initiative, location decisions have proven rather more difficult. As a result of increased land prices in suburban neighbourhoods, a number of new hospice developments have been more or less forced to establish themselves on greenfield sites in semi-rural areas which may be attractive in outlook but are poorly served by public transport. Furthermore, such isolated sites, no matter how picturesque, may unwittingly suggest that the building and its residents have been cut adrift from the wider community – put out to grass so to speak. However, this is true of other facilities for older people. Growing numbers of residential care homes and nursing homes are now located at some distance from urban centres, often for financial reasons arising from cheaper land prices and lower staff costs, though, like nineteenth-century mental hospitals, some benefits may also accrue from a quieter, more rural setting.

Nevertheless, most new hospices have successfully settled and integrated into their neighbourhoods, and public opposition by residents to a proposal to develop a hospice remains rare. Indeed many have become a source of local pride, as is evident by the degree of local fund-raising and community activities in support of their work evidenced almost everywhere they have been located. Princess Alice Hospice, already mentioned, runs 26 shops in towns and villages in its catchment area.

In order to fit into their suburban or greenfield settings, hospice buildings are often required not to draw too much attention to themselves, both for planning reasons and for reasons of preferred style. They are mostly two-storey, red-brick campus type settlements, enclosed within gardenesque landscapes, though often with what seems an excessive amount of car parking space, though this is sometimes cleverly disguised. Many hospices seem to go out of their way to organise open days and community events, including inviting local school students to visit, and make every attempt to 'normalise' a setting that is in most other ways quite fearful, or certainly would have been regarded as being so only a generation ago. Such an approach contrasts with that of most hospitals or nursing homes, where public interest or involvement is less successful. In domesticating the place of death in this way, hospices have effected a quiet revolution in recent decades which has not yet been widely registered, let alone explored.

Entrances and exits

One of the main unresolved processional issues with regard to hospice architecture relates to entrances and exits – a basic design matter for all buildings but much more symbolically so for the hospice (as it was when the crematorium was invented, in not entirely unrelated ways). Patients, visitors and staff enter through the front entrance, but bodies are discreetly taken away from the rear, usually through a double-door access bay that is either artfully contrived to look domestic in size or is shielded from public view by fencing or planting. Not everybody I talked to was happy with this overly covert arrangement, but it remains a processional issue of such symbolic importance in its wider implications that it has still not satisfactorily been resolved. The anxiety about this design aspect derives from the degree to which the death of a patient and their 'removal' – as the taking away of the body is called – are signalled or made explicit to other patients and visitors. In terms of physical removal this is still usually disguised, by discreet removal from the rear of the building (see Chapters 4 and 5).

The degree to which a death is publicly acknowledged varies from hospice to hospice, though it is rare to make a formal ritual event of the removal of the body. Sometimes a candle is lit in the foyer when a death occurs, or is announced in other ways, though not always. In more religious hospices – and most have their origins in the Christian faith as their naming after saints patently indicates – weekly services in the chapel will include prayers for people who have recently died. Hospice staff themselves often agonise over the spatial and processional taboos and boundaries associated with arrival and departure, knowing that in nursing homes things are often quite different. Residents of nursing homes have on occasions insisted that the undertaker remove a body out through the main entrance as a mark of respect. As one care home resident once told Baroness Julia Neuberger, 'We came in at the front door, and we want to go out by the front door' (Worpole, 2009: 57).

The difference between a hospice and a nursing home on this matter – and it is only one of many of course, and there is a limit to any other parallels which might be drawn – may be explained by the nature of the friendships which can develop in nursing homes between residents. In the nursing home residents may have been there for a number of years and developed close friendships. As a result, the death of any one member of such a group is an intimate event and may well be marked as a 'family affair'.

The hospice patient is rarely resident long enough to develop such relationships, particularly when occupying a single room, and a hospice death is to a degree handled more formally and institutionally, though this in no way lessens the respect and sympathy on the part of the staff. It was pointed out to me in several hospices that in contrast to a death in a hospital where the body may be handed over to mortuary staff, in hospices the nursing staff member attached to the patient continues to look after that person, even after death, until removal. Thus, as one staff member put it, 'there are only patients in hospices, never dead bodies.'

However, there are other aspects to the problematic front door/back door configuration. Where funeral flowers are brought to the hospice – as they often are, to be used as ornamental displays in the dining rooms and common areas – they arrive at the rear. They are then taken to a utility room where their funereal arrangement and symbolism are literally deconstructed and reassembled in a vernacular, domestic form. They thus move from the place of death back to the place of the living. This small detail says much about the still potent power of ritual and symbolism surrounding death and the many rituals which accompany it.

It is for similar reasons of processional ritual that some people still look unfavourably upon a crematorium committal compared with that in a church. In many cases there is only one public door to a church, and everybody enters and exits through that door – returning to the flow of life again – whereas at the crematorium the processional is a one-way street (or a conveyor belt as many critically remark upon the standard crematorium service). The differing status of front and rear doors permeates the symbolism and meaning of these processional events – especially at such critical life moments – and architects are still very cautious about suggesting any radical changes to the present common arrangements. This is very interesting. Normally in the life of a new building type sooner or later a 'signature architect' will produce something very radical in an attempt to re-draw the basic template. Aalto did this with the sanatorium as well as with the public library; Le Corbusier did it with the private house and the apartment block. Van Eyck revolutionised the idea of the orphanage, as did Duiker's innovation of the Amsterdam open-air school.

This has yet to happen with the hospice. Perhaps it never will. For while the client remains ambivalent about the degree to which a hospice should make explicit that it is a place in which people die – and as has been observed, more and more people are leaving the hospice to go home to die – the architect has to respect this ambivalence, designing a building which embodies a circulation pattern and configuration of spaces that likewise make no clear articulation of ritualistic or architectonic symbolism which one might have expected in a more religious or ideological time or culture. As one hospice director told me, every patient has to be treated differently in terms of their degree of acceptance or denial that they are dying, and so the hospice cannot be too explicit in its ethos or culture around the imminence and certainty of death: 'Some keep busy planning their funerals with great energy, whilst others simply cannot accept the reality of death.'

Hospice entrances are usually modest but welcoming. There is often a canopy to shelter people from the rain as they get in and out of cars or ambulances, and no grand steps or formal architectural entrance elements such as pillars, solid doors or vestibules to impede ease of access for frail people or those in wheelchairs. Modern hospice foyers increasingly resemble those of hotels, with large reception desks, many armchairs informally arranged around a carpeted and pastel-coloured interior, with magazines and flowers arranged on tables, complemented by a café area where visitors and patients can gather and feel a part of the flow of the place and its comings and goings. The worst entrances are those which have allowed car parking to gather round the perimeter of the building, nose to window, creating a defensive iron shield. Entrances are likely to face north. Their orientation is dictated by the fact that the rear of the main building usually contains rows of single patient rooms designed to face east or south east on to a sheltered, south-facing garden (Figure 3.4).

As I was told on more than one occasion, it is now commonly accepted that for dying people, it may be a source of comfort and even pleasure to enjoy the early morning sun brightening their room, rather than watch it decline in the late afternoon or evening. Orientation is a vital symbolic and practical aspect of the basic hospice footprint and floor-plan.

A number of hospices have chosen to locate their 'chapel' or remembrance room close to the main entrance, as this adds a distinctive architectural element to the front elevation, particularly if this takes the shape of a round tower, octagon or other pronounced form. These will usually contain stained glass windows which at night, even if only lit by

Figure 3.4 St Richard's Hospice, Worcester. Each south-facing patient room has its own garden patio.

candles, offer the outward effect of lighthouse, watchtower or beacon keeping sentinel on behalf of the settlement.

Some bereaved individuals will return to the hospice to sit in the chapel or sanctuary space, and will continue to do so for several years after the hospice death has occurred. These visits often take place at night, or even in the early hours of the morning. Thus the location of the chapel or sanctuary space next to the main entrance gives visitors the possibility of access at all times, without necessitating a journey through the inner corridors of the main building, consequently raising security issues or concerns over infection control. Through this common architectural device the hospice illuminates and keeps watch over its suburban location, particularly at night, a symbol that the work of the building proceeds apace, even when all around is dark or sleeping.

These sanctuary spaces are not, however, without their cultural or religious ambivalences either. In today's multi-cultural society they deploy a widening nomenclature. What were once unequivocally called chapels may now be called the 'sanctuary', 'remembrance room', 'prayer room', 'quiet room', 'meditation room' or some other appellation suggesting an enhanced quality of place for tranquil assembly, prayer or meditation. Nevertheless the common use of stained glass anchors them to the Christian architectural palette (as does the ubiquitous and related employment of stained glass in crematorium buildings), and continues to provide the hospice building with its somewhat sublimated religious aura, as presented to the outside world.

Many members of hospice staff have admitted that the 'chapel' element continues to pose problems in the design – let alone its subsequent use – of such spaces in hospice buildings. One told me, in a view echoed by several others, that 'they are either too big or too small.' At St Christopher's the future use of the existing large chapel is being debated. If too large, they take on the qualities of a basilica, presbytery or assembly hall, too monumental to provide the intimate sanctuary space most often required by small groups of patients and visitors. Large gatherings are in any case quite rare in the hospice calendar. However, if too small, there is insufficient room for a more formal event or religious service, whatever the denomination.

One response to this problem has been to disaggregate the notion of sanctuary space, providing smaller, more intimate seating areas and niches, within the public areas, where individuals and families can gather in relative privacy. This does not obviate the need for a clearly demarcated spiritual room, which, as has already been noted, often provides a distinctive architectural element to the main building which recognisably distinguishes the hospice from the nursing home, residential block or, indeed, the low-rise, country lodge or motel. Viewing rooms for the display of the body of the recently deceased patient to close relatives or friends are also being phased out in plans for new hospices (Figure 3.5).

In some hospices, their infrequent use means that they have been phased out. To compensate for this omission, some hospices have specified in the design brief that one of the single patient rooms should be able to be chilled quickly for such a purpose. Others rely on local undertaking companies to provide viewing rooms on their own commercial premises, off-site. As with the exclusion of a mortuary, dispensing with a dedicated viewing room signals the move away from the hospice as a place in which death is fundamentally located, to a place for living people – in which death and its immediate aftermath are increasingly handled off-stage (see Chapter 7).

The lonely room?

Almost unquestionably the principal locus of death within the hospice is still the single bed in the individual patient room. Fewer and fewer new hospice buildings have been designed with multi-bedded rooms

Figure 3.5 Viewing Room at Vidarkliniken Hospice near Jarna, Sweden.

(the word 'ward' is – in most places I visited – strictly off-limits), though some hospice trusts still insist on retaining the option of providing rooms with several beds for those who wish to share accommodation with others. The majority of patients now have their own room, usually en-suite, with sufficient space to provide a bed-settee for overnight visitors. It is also likely to be at ground-floor level, with access out on to a small private patio garden (which in turn opens out into a communal garden area). Hospice patients will spend the last few days of their life in this room: rarely less than a couple of days, rarely more than a fortnight.

The decision to universalise single-room accommodation in new hospital-building programmes – certainly in the developed world – has been principally dictated by medical reasons. In the long run, it is argued, single rooms offer greater infection control, and are therefore more cost-efficient. In the hospice the issue of infection control is not quite such a priority – but patient preferences are – and most hospice patients prefer to spend their last days in the privacy of a single room, with close friends and family on hand where these exist and are able to visit. Social critics such as Elias (1985) have raised questions as to this increased privatisation of the dying process in modern societies, believing that it raises some difficult questions for the nature of the individual psyche in the modern world. Some hospice staff expressed reservations about the drive towards providing single rooms exclusively, as they can and do recount stories of people who have greatly benefited from sharing their last days in a room with other patients, particularly of course those without close friends or family able to visit. While death in a public area was considered the ultimate humiliation for many, death in the private room may also have its own emotional costs (Lawton, 2000).

It is hard to ignore the aetiology of the single patient room, which can be traced back to the monastic cell, and its culture of penitentialism. It further blurred the line institutionally between care, custody and correction in the subsequent history of the sequestered institutional cell, something which I have explored at greater length elsewhere (Worpole, 2009). When Le Corbusier was commissioned to design a new hospital for Venice – which was never built, thankfully many now say – his space capsule designs for the individual rooms, known as the 'care unit cells', were uncompromisingly austere. He had been much taken by a painting by Carpaccio of the funeral of Saint Ursula, who was carried to her grave on a raised bed, her body 'elevated over the mundane', and so designed a high bed platform in a plain cell with no windows to look out of, though natural light was allowed in through a deeply recessed skylight above and behind the patient's head. The famous plan and cross section of the care unit was widely published in architectural journals, though historians James and Noakes rightly described it as 'a uniquely awful and inhumane environment for care'[10] (1994: 17). While hospice patients' rooms are certainly much homelier than this austere prototype, there remains an existential problem as to the singularity (and its potential attendant loneliness) of death in what are now regarded as the ideal circumstances of modern palliative care. It is likely that nursing homes will be required to follow the same route, providing only single-room accommodation, fulfiling the fears of Elias (1985) that death in the modern world may become an increasingly lonely affair.[11]

Notes

1. Saunders, Cicely, *Selected Writings 1958–2004* (Oxford: Oxford University Press, 2006).
2. Verderber, Stephen and Refuerzo, Ben J., *Innovations in Hospice Architecture* (London: Taylor and Francis, 2006), p. 15.
3. *Programme for the Good Hospice in Denmark*, Realdania Fund, February 2006, English translation published on website (www.hospiceforum.dk), January 2007.
4. Ibid.
5. Help the Hospices, *Hospice and Palliative Care Directory, United Kingdom and Ireland 2006*, edited by Daniel Ward (London, 2006), p. 157.
6. Department of Health, *End of Life Care Strategy* (London: DH, 2008), p. 23.

7. Healthcare Commission, Cited in *The Guardian*, London, 27 September 2007.
8. *British Medical Journal*, 2008: http://www.bmj.bmjjournals.com/cgi/content/full/328/7440/605/DC1.
9. Healthcare Commission Survey, Cited in *The Guardian*, London, 9 April 2008.
10. James, Paul and Noakes, Tony, *Hospital Architecture* (Harlow: Longman, 1994), p. 17. See also Sarkis, Hashim with Allard, Pablo and Hyde, Timothy, *Le Corbusier's Venice Hospital and the Mat Building Revival* (Munich: Prestel, 2001), p. 23; Degrémont, Nils, 'Palliative care and architecture: From hospital to people', *European Journal of Palliative Care*, 1998, Vol. 5, No. 4: pp. 127–129.
11. This is the principal argument made by Elias in *The Loneliness of the Dying* (London: Continuum, 1985).

4
Dying Spaces in Dying Places

Carol Komaromy

In England and Wales each year over 400,000 people of all ages die in institutions, most of them in hospitals (National Statistics, 2007). Statistics on care home deaths are difficult to obtain; but it is estimated that about 23 per cent of deaths occur there, the majority among women over 85 (National Statistics, 2007). Bearing these figures in mind, the care home industry itself is big business with admissions set to increase alongside the growth in the number of older people in the UK population (Laing and Buisson, 2006). The profile of residents in care homes reflects this ageing population in that people who are admitted to these institutions are likely to be older and frailer than in the past (Komaromy, 2002; Sidell and Komaromy, 2003). What is more, between one quarter and a third of care home residents die each year; in the UK care home deaths currently stand at around 92,000 a year (Mozley *et al.*, 2004).

Care home staff are required to manage these deaths, meeting the complex needs of residents undergoing physical and emotional deterioration at the end of life alongside the ongoing care of other residents' slowly disintegrating bodies. The extent to which these tasks are institutionalised forms the crux of this chapter. Building on research findings made public in the 1960s by, for example, Townsend (1962) and Goffman (1961), about the dangers of institutional life, this chapter draws on sources such as Goffman's (1961) argument that the main features of social life which produce and sustain people's identity are lost when they enter 'total' institutions. In such settings social arrangements regulate, under one roof and according to 'one rational plan, all spheres of institutional life' (1961: 18). According to Goffman, people suffer a series of mortifications of the self on admission to a 'total' institution. Ultimately, Goffman saw such processes as a rite of passage into a world in which practices of social control and restriction removed people's self-determination, autonomy and 'adult competency'. Goffman's (1963b) notions of 'spoiled identity' and stigma, which ensue as a result of these mortifications, have implications not only for life but also for death within residential care. Kellehear (2007), for example, draws on them to argue that contemporary western societies have seen the birth of 'shameful death' in which older people are hidden away and infantilised.

However, these perspectives have had unanticipated policy implications. Since the 1960s, concerns about the extent to which people in institutional care suffer the consequences of a loss of autonomy, and experience a corresponding sense of dependency, have resulted in the introduction of inspection and regulatory safeguards in UK care institutions. These are designed to avoid 'induced dependency' (Booth, 1985) and have had both direct and unanticipated outcomes for the culture of care homes of older people. In this chapter, I argue that tensions between the policies designed to mitigate the series of losses that result from institutional life described by Goffman, and the reality that residents are likely to be much closer to death as a consequence of a later admission to a care home, make the purpose of the care home as an institution markedly ambiguous.

What is more, given the profile of many residents in care homes, the focus on 'rehabilitation' which has accompanied inspection and regulatory safeguards has lent a particular status to the category of 'living' for care home residents, one where being seen to be 'living' overshadows the embodied presence of dying (Komaromy, 2009). In other words, as the findings discussed here highlight, despite the reality that many of the residents in care homes are dying, there is an emphasis on *rehabilitative* care. Thus the care home context within which 'dying' is interpreted works to produce 'death and dying' as the oppositional category of 'living' (Sudnow, 1967). As this chapter will show, one of the consequences of this is that the notion of a dying phase is narrowly defined and thus impacts negatively on the quality of end-of-life care, in part due to the short time in which this care can be delivered. In practice, therefore, there is a tension between care home's rehabilitative goals and the way that living and dying are played out and this manifests in a material sense in the physical spaces of the care homes.

This chapter therefore considers the way that dying spaces are created in the settings introduced above, including the removal of the dead body from the home. This is followed by a discussion of sudden and unexpected death and concludes with an overview of what might be taking place, symbolically, in terms of how staff manage death and dying in the care home and the purposes served by this management.

The study

The data used in this chapter come from a study commissioned by the Department of Health as an investigation into the case for palliative care for older people in care homes in England (see Sidell *et al.*, 1997). I was employed as the full-time research fellow on the project and used part of the ethnographic observational data for my doctorate. This comprised interviews and observations made over a period of 1 year in 8 care homes. Here, I am drawing on data from two of these homes to highlight key aspects of living and dying spaces.

Observing everyday life in care homes for older people revealed the strategies that staff and residents deployed to manage 'death and dying' and the complexities of policing what I came to understand as the narrow margin between life and death. When I talked to care home managers they were often clear that they wanted residents to have a 'good death' (Neuberger, 2004). For them, features of a good death included its predictability, with relatives being made aware of the imminence of death, and its timeliness, not just in terms of age but also in terms of the duration of the dying period. The reputation of the head of a care home could indeed rest partly upon the accuracy of such predictions. I heard accounts of how those residents who were prematurely categorised as dying or, conversely, took several months after a prediction to actually die could present problems, both in terms of the resources needed and the embarrassment associated with getting the prediction wrong. Therefore, among other purposes, delaying a diagnosis of dying increased the chances of certainty.

Therefore, in this chapter I argue that institutional practices framed the physical and emotional spaces within which dying and death in the care home were produced and experienced, with dying largely confined to a narrow temporal margin very close to the end of life. However, the overwhelming focus on physical care orientated towards 'rehabilitation', combined with the late, and sometimes arbitrary, diagnosis of dying, meant that predictions were not always accurate. For those residents who were experiencing a slow and gradual decline to death, establishing the beginning of 'dying' was difficult, yet having a distinct form of dying that was fixed and certain emerged in the study as one way of potentially maintaining a boundary between living and dying residents.

Theoretical resources

Anthropological theories have drawn attention to the importance of social boundaries of the kind described here: van Gennep ([1909] 1960), for example, argued that, particularly at birth, puberty, marriage and death, rites and customs serve important psychological, sociological and symbolic functions by not only marking but also *making* changes in the ways in which individuals are positioned within society. Within his three-part rites of passage schema, he was able to subsume the complexity of observed ritual practices across many societies. Specifically, he argued that through ageing and illness individuals withdraw from active life and social contact in ways that carry implications for their social identities. Van Gennep also generated the concept of liminality to describe the liminal or transitional phase which is central to any rite of passage and is likely to be marked by ambiguities of many kinds. Subsequent anthropological work on boundaries or margins carried out by Douglas (1966) argued that they are both invested with power *and* are potentially dangerous, requiring some form of (ritual or symbolic) management. These perspectives help make sense of the way in which maintenance of the boundary between life and death was managed or 'pulled off' by appropriate performances within the care home, ones which at the same time serve to 'mask' or create distance from a death. I argue, for example, that a focus on the physical care of dying residents and on the physical aspects of the 'living' body, described later in this chapter, may serve as this mask.

Douglas' (1966) concern with the symbolic significance of boundaries thus helps explain the need felt within the care home to keep 'living' and 'dying' categories distinct, a concern that she elaborated through the concepts of pollution and taboo. Thus, she argued, rules of hygiene are only partially concerned with the biomedical dangers associated with dirt; instead, what is commonly categorised as dirt can be seen as 'matter out of place', that is, matter which flouts a society's classificatory system and occupies a state of categorical ambiguity. Practical responses to 'dirt' and its associated danger of pollution can often be understood to have ritual or symbolic, rather than purely instrumental functions. Thus ageing, decaying, leaking, smelly and confused bodies might comprise what Douglas refers to as 'matter out of place' in that they offer resistance to any tidy division or boundary between life and death and living and dying. Indeed, death itself has long been perceived as dirty and polluting for, as Lock suggests, 'Whatever form death takes, it conjures up that margin between culture and nature where mortality must be confronted' (1996: 235).

Drawing on the theoretical resources provided by sociologists such as Goffman (1961) and anthropologists such as Douglas (1966), therefore, I now consider the following ethnographic material which details the materiality of death in care home settings.

The creation and production of living spaces in care homes

The first care home I consider is Regis Nursing Home.[1] Situated in the West Midlands and housing 36 residents, Regis home was formerly a domestic home before being converted and extended into a nursing home. It was in this home that I observed the way in which 'living' was produced, and via data deriving from this home I will also detail the death of one of the residents, the part played by the layout of the home in her death and the subsequent removal of her body from the home.

In this home, I recorded the work of the activities helper (as she was called) who provided occupational therapy for some of the residents. This role was not uncommon in many of the care homes in which I made observations and was part of meeting the minimum standards of care. Activities for residents, created and facilitated in the public spaces of the care home by the activities helper, provided a highly visible form of rehabilitation that operated to present residents as 'living'. However, this performance of 'living' was problematic, as illustrated below:

Field Notes: Regis Nursing Home

> *Maude sat at the table with her back to the lounge and the other residents with a jigsaw in front of her. Some of the other residents in the room were decorating hats and this is where June, the activities helper, focused her attention. Maude stared out of the window. I was seated just to the right hand side of her a few feet away. The jigsaw was on a tray and about a quarter of the pieces were in place and the rest scattered around the edges of the tray. Every now and again Maude let out a deep, audible sigh. This seemed to remind June to come and look at the jigsaw and ask Maude how she was getting on, (which she did three times during the hour set aside for activities). On one occasion when June was with her, Maude looked at the jigsaw and picked up a piece and stared at it. 'That's it – you see if that*

> *goes somewhere', said June. As soon as she walked away, Maude stared out of the window again and let the piece fall out of her hand onto the tray. I approached Maude and asked her if she liked jigsaws. 'I hate them', she said and let out another loud sigh.*

From her appearance seated in front of the jigsaw it might be interpreted that Maude was 'doing a jigsaw'. As this extract from my fieldnotes indicates, however, the reality was somewhat different. Maude was immobile and had to be helped to move from her wheelchair into an armchair, to the toilet and into bed. 'Trapped' in front of the jigsaw, it did not seem to matter to the activities' officer, at least, that Maude did not enjoy 'doing' it, nor that she was unable to place any of the pieces. What seemed to be more important was her *appearance* of doing the jigsaw – of 'living' – and how this contributed to the activity session which formed part of the overall performance of rehabilitation in Regis Nursing Home. However, rather than reducing dependency, this performance of rehabilitating Maude was itself in danger of becoming simply part of an institutionalised form of living; it was a routine event that did not require Maude to be involved in any meaningful way, other than to demonstrate the actions of being involved, a surrogate for the activity of rehabilitation.

From my observations and conversations with staff and residents, then, keeping immobile residents mentally occupied seemed to be the most important aspect of home activity specialists' work, in contrast with the physical care that dominated the routine and relentless work of the home care nurses and assistants. In this way the separation of the mental and physical care of residents reflected an artificial distinction between their minds and bodies. Despite the significance of what Goffman (1961) would call the loss of identity (and what Sudnow, 1967, calls 'social death'), I argue that in Regis, it was primarily the appearance of 'mental activity' which was introduced as a supplement to physical independence that allowed residents to be seen as conforming to the activity of 'living'.

The second home I consider in this chapter was Poplar Court, another nursing home in the West Midlands which accommodated 32 residents in need of nursing care. Unlike Regis, this home offered purpose built, single-storey accommodation and here too I explored the activities that contributed to the production of living, as opposed to dying. These took place in the communal spaces of the home which were open to public visitors and not designated as a private space, such as the residents' bedrooms (Figure 4.1).

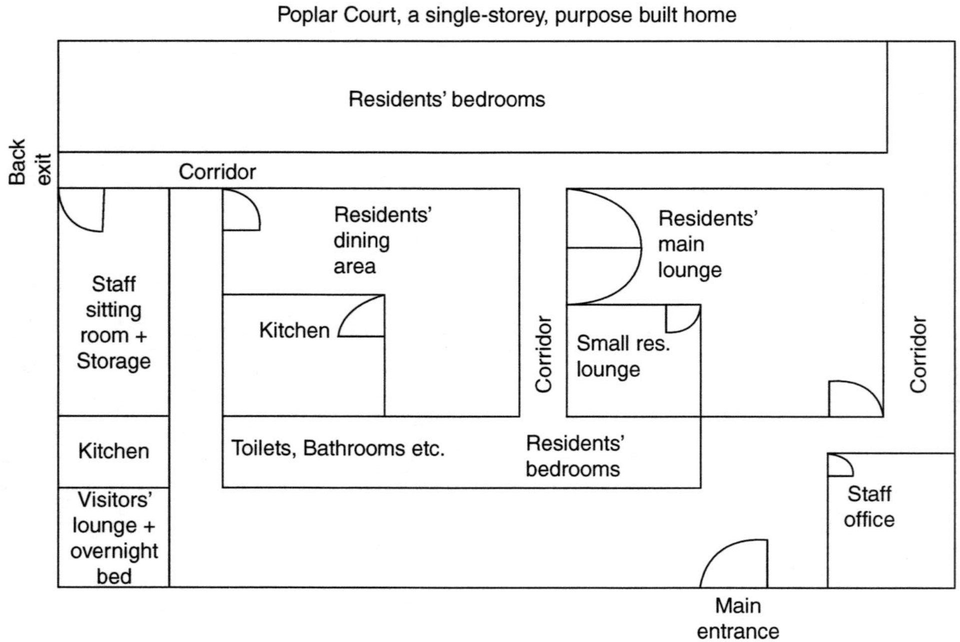

Figure 4.1 The layout of Poplar Court highlights how the home was designed so that the 'living' areas of the home were at the centre – with the bedrooms and utility rooms around the periphery.

As in Regis, at Poplar Court, where residents' levels of physical dependency were high, the performance of living was produced through home routines.

Extract from field notes at Poplar Court

At 11.30 the staff began to move the residents into the dining room in wheelchairs to tables for their lunch at midday, while other residents were helped to their feet and made the slow journey by using a Zimmer frame with some assistance from a staff member. The television set was turned down and the noise of shouted instructions replaced its sounds. 'Come on Mary; put your foot forward, that's it'. 'No!' shouted another care assistant, 'you have to go to the toilet, before your dinner!' The nurse in charge of the shift gave me my instructions. 'Can you wheel this one in please and you can feed her.' This 'one' was Daisy, a large woman in a wheelchair and one of only two black women in the home. She did not speak to me and had the lopsided facial appearance of someone who had suffered a stroke.

Other residents were wheeled to a table and remained seated in their wheelchairs. All of the seven residents who needed to be fed were seated close to each other. The kitchen staff appeared with meals on plates and the staff put 'bibs' on some residents while others had serviettes. I was enlisted to help feed more residents, after I had fed Daisy, who chewed her food a lot and did not seem to want to swallow. I felt anxious that I was slower at feeding than the care assistants who were cramming the food into residents' mouths. I noticed that their residents were willing to be fed. They ate quickly and sat with open mouths waiting for the next mouthful (like baby birds in a nest). At last, I had fed my charges and there were only a few residents left at the table.

Unlike Regis, at Poplar Court home staff focused almost exclusively on the *physical* side of caring, ensuring that the residents in their care looked as well as they could; a condition I interpreted as 'appearing to be living'. The two public spaces at Poplar Court were located in the centre of the building and as such were key to the routine activity of sustaining 'living'. Furthermore, through unrelenting care practices the residents' *physical* bodies were symbolically highlighted or kept to the fore as the primary representation of this category. For example, immediately after breakfast, staff washed and dressed all of the residents and then wheeled them from their bedrooms to the sitting room. The contrast between the movements of the staff and the stillness of the increasing collection of seated residents was striking.

The key practice of living that residents in Poplar Court had to achieve was that of eating and drinking, and the routines of the day were organised around mealtimes. For example, after lunch was served and 'eaten' (and again in common with many homes), the exodus to various parts of the home began. Some of the residents were taken to the toilet or, as staff put it, 'toileted', a term that transformed a practice normally undertaken independently by adults into something *done* by staff to residents. This example demonstrates the ambiguous nature of a setting oriented towards rehabilitation which produced 'living' not by enabling the practices of everyday life but by doing them *to* people. After 'toileting', residents were returned either to their bedroom for an afternoon rest, or to the lounge area. At 5 p.m., they were given tea which involved further migrations to the dining area; then most residents were taken to bed, a procedure which overall took over 2 hours. Throughout these tasks, the staff wheeled equipment, commodes, hoists, wheelchairs, to and fro, sometimes containing a resident and sometimes not. These activities helped (re)confirm that the resident was living on the basis that they were, in some way and with varying degrees of support, physically mobile. Importantly, then, by participating in 'living', via mealtimes at Popular Court, for example, residents were categorised as not dying.

This assertion resonates with Sudnow's (1967) argument that death is both natural *and* social. To avert each of these forms of death, physical activities such as eating were institutionalised, thereby feeding residents' bodies, to some extent, *and*, importantly, affirming their membership of the *social* category 'living'. Thus, food and drink were media through which people could acquire and assert a living status; they both sustained the 'natural' living body *and* were cultural symbols of life. Even for those residents who struggled with eating, being with others at mealtimes, taking food into their mouths and participating in activities such as occupational therapy, could be re-presented as visible performances of the tasks of living.

Data also revealed that residents occupied various *states* of living so that for some residents living was easier to achieve than for others. What I mean by this is that there were no objective measures for establishing how *well* someone was living. For example, living could be loosely defined as not dying and required from high to low levels of care needs. Indeed, 'living' was often itself a performance fulfilling the appearance of social connection, rather than a form of meaningful care in which the residents felt sustained and valued. Furthermore, an emphasis on *rehabilitative* living in both homes left little room for those residents in need of end-of-life care; as a result one of the first signs for other residents that someone was ill and possibly 'dying' was when he or she did not appear in the public spaces of the home where rehabilitation took place. Unsurprisingly then, removing a dying resident from these public spaces was extremely significant for other residents.

Dying in the institutional space of a care home

As my fieldwork observations revealed, the extent to which the boundaries between life and death were maintained in both care homes and the strategies that staff deployed in order to do so were variable and indeed ultimately limited, given the intractability of bodily deterioration. One such strategy involved separating dying residents from those who were living, the first significant change in the routines and practices of care homes that marked the beginning of the dying process (Figure 4.2).

As the following account of the last few hours of a resident's life indicates, the way that the physical space of the dying resident's bedroom was used and staff's commitment to keeping dying physically concealed from other residents represented one of the home's strategies for sustaining the social category 'living'. Yet, as the account shows, this resulted in inherently contradictory performances around dying and living residents.

Alice's withdrawal from public space: Extract from fieldnotes

> *Matron had drawn up a new care plan. Alice's bed was against the wall and the staff had to pull it out when they routinely turned her to the other side, which they did every two hours. Alice had difficulty in swallowing and this frustrated the staff's attempts to give her fluids, even through a syringe. The prescribed pain relief was likewise difficult to administer. The staff did not sit with Alice, although they expressed an intention and a desire to do so. Instead they visited her every 15 to 20 minutes.*
>
> *When I talked to the residents in the lounge in the part of the home in which Alice sat most days, they expressed a lot of concern and shock at the 'suddenness' of her illness. Martha in particular was tearful and told me she was, 'praying that she lives, and doesn't leave me'.*
>
> *(...)*

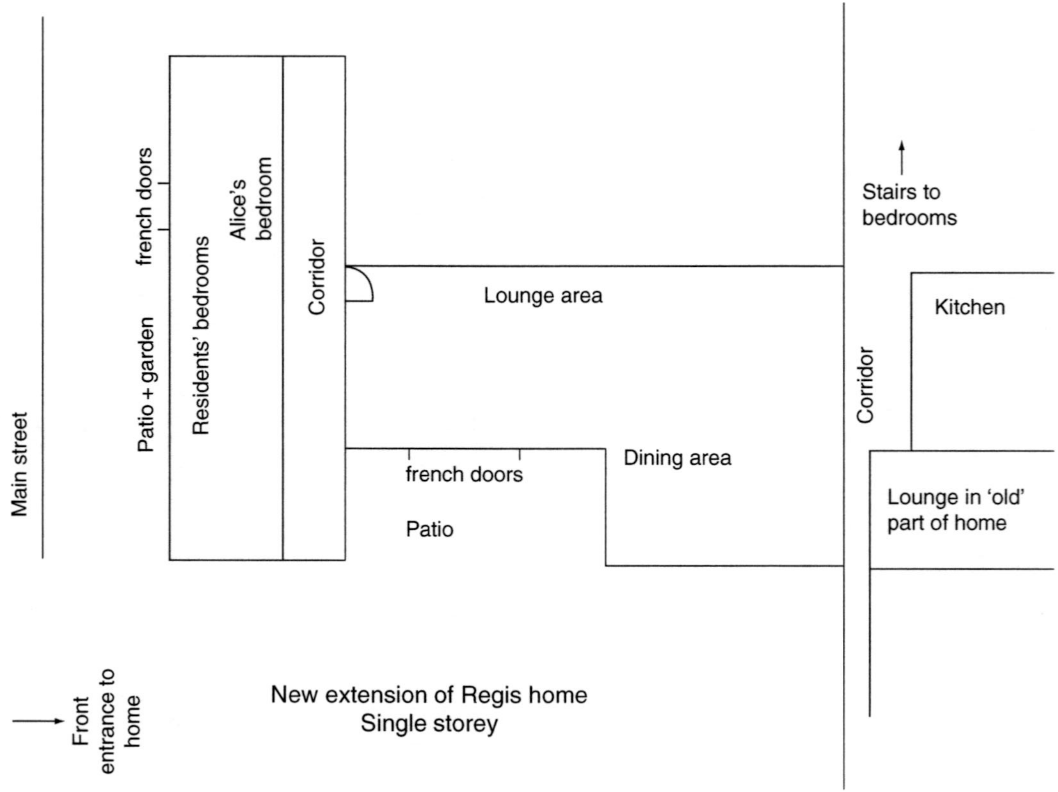

Figure 4.2 This simple plan represents the bedroom area of the new wing of the home and its geographical location in relation to one of the lounge areas, also in the new part of the home. The bedrooms provided an exit through the french doors through which funeral staff removed bodies.

> *Each time I passed through the lounge I noted how the remaining residents continued to sit in their usual place, around Alice's empty chair.*
>
> *When I visited the next day and talked to matron, she told me that there had been a slow deterioration during the night and Alice's breathing was more laboured. She called this the 'death rattle'. I went to see Alice, and, although she was less responsive than the day before, she appeared, at times, very agitated and still screamed out when she was moved, even though a care assistant told me she was 'more peaceful'.*
>
> *Alice was given Oramorph*[2]*, by mouth through a syringe and in small and frequent doses after she was turned, rather than before. Apart from the GP's daily visits no other specialist professionals were called in.*

The head of home told me that Alice had lived in the home for 12 years and made new residents welcome by befriending them and helping them to settle in. Once her death was predicted, Alice was separated from the other residents, some of whom were close friends. This was achieved not just by keeping her in bed in the concealed space of her room but also by the lack of access that was afforded to those friends. During my fieldwork, and when Alice was 'living', she had been the focus of her friends' attention and I was struck by the way in which her empty chair in the lounge provided a powerful symbol and reminder of her absence from this 'living' space. In part this absence served to signal her 'dying', and the residents in the lounge who were close to Alice did not seem to want to collude in this concealment of her dying condition and frequently asked me how she was.

While my data suggest that the practice of separating living and dying residents by moving dying residents into sick bays had largely ended (Hockey, 1990), living and dying residents continued to be kept spatially apart. This separation was one of the first responses to residents' 'dying', a process that often took the form of a spiral of gradual but uneven decline among the frailest residents in both homes. As a result, accurate predictions about exactly who was dying were difficult to make. For example, a resident identified as dying could sometimes remain alive for much longer than expected. Despite this, the strategy of separation meant that once a resident crossed the living and dying boundary, it was difficult for them to return to living space. Instead, he or she was suspended, either in a 'state of dying' over a longer period than anticipated, or as van Gennep ([1909] 1960) might have argued, in an ambiguous liminal space between living and dying, one which brought social if not bodily death. Scope for occupying a liminal period could, however, extend beyond bodily death when residents' bedrooms became the container of their bodies which, whilst 'dead' in a physical sense, required the social confirmation of a death certificate that could only be conferred by a medical professional rather than a member of care home staff.

Once this had occurred, however, the dead body became the 'material reality of death', and its disposal offers significant clues about its new identity as a corpse (see Hallam *et al.*, 1999). To explore this process, I draw on my direct observations of what happened to Alice after her death.

Removing Alice from the home: Extract from fieldnotes

> *At 6.15 pm, shortly after being called, the undertakers arrived at Regis House to remove Alice's body. Matron and I went with them and on the way through the lounge to the bedroom matron closed the curtains in the dining room, so that the residents would not see the hearse. I helped matron to move furniture around so that the undertakers could access the French doors. The room was very small and making room for the mortuary trolley was difficult. At one point, the commode that was in the room toppled over onto the dressing table and made a loud crash. I wondered how much of this noise the surviving residents were able to hear and what they thought we were doing. The undertakers took Alice's body out in a body bag on a trolley through the French doors of her bedroom unseen by other residents, who were still having tea as part of the normal routine.*

Even as the inevitable product of many homes, the residual dead body lying in a former resident's room was still perceived to be problematic by care home staff in that it had somehow to be moved across the home's 'living' space, potentially frightening those residents who still, somehow, adhered to this status. Its exit therefore had to be contained and concealed. Yet, as with residents whose absence from living space marked their dying, efforts to conceal the body's exit commonly served to *expose* what had taken place (Komaromy, 2000). Despite the fact that Alice's death was anticipated and that the care staff considered it (in retrospect) to have been a 'good death', the removal of her covered body from the home constituted a physical reminder of death and so became something that had to be concealed. The surviving residents were considered by Matron to be in need of protection from the sight of her body leaving the home, even though Matron announced her death to them shortly afterwards. Importantly, the difficulty of concealing

Alice's dead body underscored the fragility of the boundary between life and death.

So far the data presented have demonstrated the considerable effort made by care staff to separate living and dying residents, a strategy which included careful predictions about the imminence of death, and attempts after death to conceal the body of a deceased resident. Dying thus took place in a tightly orchestrated way. The processes through which the news of dying and death was shared with other residents – such as Matron announcing it to the residents – reflected the home staff's attitude about death more than the needs of individual residents. For example, data from one of the other homes in the study, a Catholic home, revealed that news about death and dying was shared more openly with other residents than in the case at Regis (even though the period of dying was still confined to days and weeks). However, this policy of open sharing was based on the assumption that all residents would want to be included in the news about someone's death; some residents told me they did not want to know when someone had died.

Sudden death and the ruptured performance

In the case of an unexpected death, all of the strategies described for maintaining the boundary between life and death were threatened. My data showed that deaths of this kind were considered to be (matter) out of place (Douglas, 1966), both physically and temporally. As such they threw into chaos the ability of care staff to carry off the performance of a good death scenario. However, rather than just a sense of having failed to achieve a good (i.e. predictable) death, sudden and unplanned deaths carried much more significance and threat for care staff and other care home residents. The discovery of a dead resident had the potential to erode the institutional control of death maintained through the separation of living from dying residents by exposing the care staff's ultimate failure to sustain this arrangement. In this way, a sudden death threatened the authenticity of the performance of managing the end of life and producing a good death.

From the accounts of sudden death that care staff shared with me in all homes in the wider study, the degree of shock that resulted from finding a resident suddenly dead in a space that was preserved for living residents transgressed the protocols of what living spaces in the home were used for. When death forced itself into space usually protected by the *removal* of a dying resident, this created chaos in the home. I have argued elsewhere (Komaromy, 2002) that in terms of a hierarchy of difficulty, the least problematic space in which to cope with sudden death was where it occurred in a private space in the home because it could be concealed; the most problematic was one where living residents witnessed the discovery.

Those staff who talked about encountering sudden death told me of their own distress, something mitigated only by the ease with which they were able to conceal the dead body. For example, one carer told me that she had lifted a day-care resident into a wheelchair to convince nearby residents that the resident was still living and in need of medical help. As with Maude in Regis Home, her action made the dead resident 'live'. Furthermore, although it was not unusual in the case of anticipated death for a deceased resident to be kept in the home overnight, if there was a sudden death, then keeping the resident's dead body until a doctor arrived either to certify the death or request a coroner's enquiry seemed to cause a much higher degree of discomfort.

Dying out of place in an institution where staff controlled every aspect of a resident's life meant that this had occurred 'without permission', so contravening the implicit or informal rules of the institution (Goffman, 1961). Not having protocols in place for dealing with this event served the purpose of continuing to outlaw sudden death, a stance that reflected its undesirability. Furthermore, this lack of accommodation sustained the unacceptability of an unpredictable event in institutions where the premium was on shoring up and clarifying potentially blurred and fragile boundaries between life and death. Accommodating sudden and unexpected death through routines and procedures would threaten existing productions of living and dying and might lead to total chaos.

Significantly, the manifestation of a suddenly dead body revealed the arbitrariness of the categories 'living' and 'dying', an event that resonated with the reality of the imminence of death for many of the older residents. In many homes care staff members told me that, while they thought that residents knew that death was close and that the death of other residents served as constant reminders, they did not want to 'stuff it under their noses', as one head of home expressed it. Some residents told me how shocked they felt by the sudden death of other residents; even though a significant number confided that they liked the idea of dying suddenly (in their sleep) and not enduring a period of suffering prior to death. Despite this, at the time of sudden death it was as if all of the systems that were deployed to make something difficult more manageable and provide a particular meaning were thrown into chaos and thus were in danger of being revealed as arbitrary.

Conclusion

This chapter has argued that care home institutions managed older bodies in space and over time as part of institutional routines and roles which constituted a form of control. However, my data showed that – rather than being fixed and top down, as Goffman's (1961) view of total institutions suggests – this control was inevitably fragile. My data showed both staff and residents' failing to adhere to, or being unable to conform to, a spatialised system of control. Clearly these were not deliberate acts of resistance (Foucault, 1977), but instead challenges to the routines imposed by the institutional care home system. Foucault's argument that the body is something that has been produced, rather than suppressed or controlled (1977), allows for an appreciation of such forms of resistance and this is what my data indicated. For example, as I have shown, an ageing body is not without agency and can simply fail to perform all of the tasks of living, dying and death. Thus Maude might have given the appearance of doing a jigsaw but was not in any way meaningfully involved in the act of being rehabilitated. Just as putting food into the mouths of residents who failed to swallow it was symbolic rather than instrumental, so practices and routines within care homes themselves failed to accommodate certain types of death: sudden and unexpected deaths, and deaths 'out of place' illustrate these extremes of resistance. Therefore, I would argue that while institutions might be totalising, Goffman's (1961) account of the effects of institutions on their inmates, as simply a negative and repressive force, was not entirely helpful. Indeed, I did not find this to be the case and rather, would argue that the use of control (and power) in care homes for older people was much more productive, nuanced and diffuse.

Instead of recognising end-of-life care needs for many residents, there seemed to be an increased effort to order the dying spaces for living and dying in such a way that narrowly defined 'good' deaths could be achieved. Therefore, I argue that all these aspects of performance, control and resistance were played out with varying degrees of ease and discomfort. Just as the ageing and dying body unravelled on its way towards death, so too forces and constraints which both shaped and responded to its management were entangled and messy. With the entire focus of the practice of care homes on rehabilitative living, followed by a narrowly defined period of dying, this would explain why there was no script for any other form of death and how those deaths which were more achievable as 'good' were only possible if the period of dying was not too long or complicated.

Notes

1. The names of homes, residents and staff have been changed.
2. This is the recommended form of pain relief and believed to be more effective than a larger dose that is given less frequently.

5
The Materialities of Absence after Stillbirth: Historical Perspectives

Jan Bleyen

This chapter explores the way places and objects impact on the everyday lives of parents whose baby died at birth,[1] examining embodied actions which involve 'stuff' connected to the deceased baby. More specifically, it asks how concepts like 'death', 'grief' and the 'life course' are performed and what role metaphor might play in making sense of this type of loss.

This chapter draws on data from 45 open-ended interviews with Flemish parents (mostly women, sometimes couples) and approximately 60 birth professionals and engages with the history of not only the material culture surrounding stillbirth, but also the senses, emotionality and identity as they are affected by the death of a newborn baby. Here I use two particular accounts of the experience of stillbirth in hospital, from 1966 and from 2003, to provide themes about the changing cultural environments of stillbirth.

I therefore explore the consequences if the dead baby had no tangible, embodied social identity, a status which meant that for their parents it was unclear who and what has been lost. I will also show how objects, such as the physical body, the grave and the cradle, could all be utilised by parents in different ways and in particular material contexts – such as the hospital, the cemetery and the home – to sustain an ongoing relationship with their lost baby. In other words, this study of stillbirth in Flanders supports the argument that bereavement is, as Hockey *et al.* describe, 'a spatialised experience, an event and process located in a particular landscape' (2001: 744).

Metaphors

The processes through which the materialities of loss take on particular social and emotional meanings and the ways in which they position both living and deceased people in social space can be explained via the concepts of metaphor and metonymy. Drawing on the notion that '[m]etaphor is, at its simplest, a way of proceeding from the known to the unknown' (Nisbet, 1969, cited in Turner, 1975: 25), this chapter illustrates how the 'unknown', often 'unimaginable' event of death at the beginning of life can be made 'knowable' by and for parents through embodied engagement with the materialities of the deceased baby's body, the grave and the cradle. The power of metaphor emerges both through the parent's actions and through their metonymic relationship with the imagined life-to-be of their dead child. In this way, death, grief and loss can be encountered and made sense of through materially grounded metaphors (Hockey, 1990). For instance, as shown here, the therapeutic imperative to 'let go' of the dead person in order to be 'healed' of grief (Klass *et al.*, 1996) materialized in the embodied act of dismantling a cradle, a practice which created an empty and provocative space in the living room and reinforced the baby's absence. Thus, metaphors are not merely decorative features of language. Their efficacy emerges through their embodiment in tangible entities where they then inform practices within material contexts (Fernandez, 1972), making the abstract knowable and, potentially, sufferable (Hallam and Hockey, 2001).

Therefore, in order to understand meaning-making as a *process* constituted through cultural practice within social interaction, the chapter also draws upon the idea of performativity (see Roach, 1996). Occurring within space(s), such as the hospital, the cemetery and the home, and in relation to the materiality of objects within those spaces, such practices point towards a notion of agency which transcends human intentionality to incorporate the effects of the non-human world of space, place and objects (see Latour, [1983]1993). From this perspective, interpretations of death and grief, as well as other notions – for instance, 'father', 'shame' or 'God' – are not 'fixed'; rather they are continually co-created through the performative interactions of individuals (see Burke on the 'performative turn' in history, 2005), operating within spaces and through objects. This notion of performativity resonates strongly with the idea of 'intersubjectivity', where it is in embodied social interaction that meaning is created, enacted and sustained (Rapport and Overing, 2007).

In addition to the events surrounding stillbirths, the open-ended interview which I used as my principle research method can be interpreted historically from a performative viewpoint. During their encounter with me, narrators re-enacted their past in an embodied manner. Sometimes they brought in specific objects (such as a birth certificate or an album) in order to talk about particular memories. Just as these parents of a stillborn baby needed embodied actions to make their child real and present (as illustrated later in this chapter), in interviews historians can counter the absence of the past by making use of both *the archive* (concrete texts and objects that are kept in a physical space, in this case: the informal sphere of the private house) and *the repertoire* (embodied memory conveyed in gestures, spoken word and other ephemeral performances, in this case the narration of their experiences (Taylor, 2003)).

The 1960s versus today

Until recently, stillbirth was neglected by historians, receiving attention primarily within psychological and therapeutic literatures which focused on the bereaved individual rather than on their social and cultural contexts (Layne, 2003: 22). As a result the ways in which the notion and implications of 'stillbirth' might vary and change, both historically and cross-culturally, have typically been overlooked (Cecil, 1996).

This chapter provides an analysis of such change in Flanders, Belgium. Using a historical perspective allows me to narrate cultural shifts in how parents experienced the absence of their child and how this shaped their ongoing relationships. Similarities and differences between two women's accounts, in 1966 and 2003, demonstrate how the material and metaphorical context of stillbirth can shift. These two examples will be supported by other examples from my research in Flanders.

The first example is from Anne's personal story which provides a social and cultural narrative. One year after her wedding in 1966, Anne's first baby died at birth with spina bifida. In her interview, she described a world that *was* – albeit one that had only recently gone. For example, when their baby was stillborn, Anne, a teacher, and Francis, her husband, an academic, had not seen any ultrasound images of him. This technology had yet to figure within routine antenatal care. Because of this, their baby lacked a visible and embodied identity separate from his mother's body.

As well as technological change, Anne's story reflects other shifts in expertise and knowledge. When her baby died in the 1960s, doctors were the figure in authority at the time of both the birth and the death, dictating what happened to their patients. Today individuals usually have more choice about what happens to them (see Walter, 1999; Bleyen, 2005). However, shifts in the authority of religion were already evident at that time. For example, when a nun told Anne that '*Jesus had his purposes with the death of her child*', she stopped going to Sunday Mass, not because she was no longer a religious person, but rather because she was reluctant to accept church authority. In the wider society too – in Catholic Flanders – secularization was increasing (Brown, 2001). Anne's embodied action – staying at home instead of attending church as expected – thus further illustrates how autonomy and self-expression were emerging as new dominant values at that time.

Anne's interview data also show the resonance of historically specific metaphors. Indeed, in the course of her interview she moved from one metaphoric framing of her experience to another: the 'bad dream' metaphor she used to describe painful hospital experience changed into a 'walk' which enabled her to reclaim agency, to put things right and to turn pain into consolation. In other words, her narrative represented her mourning process as a historical journey, a metaphor which was then given substance in her changed orientation towards what had happened: from being passively at odds with the death of her son to an active subject able to relate at a distance to his loss (Bleyen, 2009b).

In contrast, Sarah, a 30-year-old therapist whose first baby died in 2003, referred to experiences and practices that were familiar to her contemporaries such as me; for example, her account echoed the current popularity of psychological rather than religious discourse. So while for Anne the interview was a chance to 'confess' on behalf of her deceased husband the acts he regretted, for Sarah, the interaction provided a potential 'therapeutic' moment. Thus, as the interviewer I was given multiple roles throughout the research process, not least as the receiver of confessions and as a therapist. In my attributed role as therapist, Sarah was noticeably more comfortable than Anne in talking to me about the stillbirth, although this may also be because she and her partner frequently spoke, with self-awareness, about their loss.

Another major difference between these two interviews was the length of time since the stillbirth. Sarah described a stillbirth that occurred a couple of years prior to the interview in February 2006; Anne's stillbirth happened four decades earlier. Sarah's account lasted three times as long, her story being far less condensed than Anne's narrative, even though this had its own history, her past having been shaped and reshaped several times during the intervening period. Significantly, Sarah also had many more objects available with which to tell her story, such as photograph albums, whereas Anne, like other interviewees of her generation, showed me the only thing she had: the marriage certificate that referred to the 'birth of a lifeless infant of the female sex'.[2]

The hospital

Like sudden deaths among older adults living in care homes, death in the delivery room also constitutes 'matter out of place' (Douglas, 1966; Bleyen, 2009a; see also Chapters 2 and 4). New life is the expected outcome of birth. When this does not happen, data indicated various strategies utilised in the delivery room to manage the stillborn baby's body. That said, how the baby's body is treated has varied historically.

For example, until the 1980s, in Flanders, birth professionals routinely hid the baby's body in a hospital side room. Current policy advocates that parents see and hold their son or daughter. This raises the question of what the actions of hiding and showing do to the baby, whose body will eventually disappear permanently from its parents' sight, never growing up, never ageing. In both scenarios – hiding and showing – the baby's body is being treated within a dynamic context of social action. Historical variation in how this body is identified can be summarized as a shift from being a distant 'it' or 'physical body', to becoming a 'social body' or, potentially, a living person with an intimate name. Only recently have parents in Flanders been encouraged to undertake 'parenting' practices such as care of the body of their stillborn baby. The result of this is that the person-that-never-would-be is being made to 'live' in a metaphorical sense in that practices which normally surround the 'known', anticipated birth of a live baby are transposed to the corpse of a dead one, so enlivening it in a social sense.

'It'

Until the 1980s, however, for many parents their dead baby remained a vague reality. Removed to a hospital side room, the 'property' of birth professionals, their child became a dead body, an impersonal 'it', rather than a personalised 'he' or 'she'. That said, for some parents the body meant so much more (see Mol, 2002). While the hidden body might not stand in for the person the child could have become, for some parents it could still constitute more than just a symbol of their child (Van Reusel, 1999: 36; Bleyen, 2005: 136). Both Anne and Sarah wanted their child's body treated respectfully. Having participated in its interuterine life, they knew it as an intimate being, unlike the doctors who surveyed it for medical facts. For them, potentially, their baby's body had a social character (see Hallam *et al.*, 1999, Chapter 7) interwoven with their own character and their child's persona. However, as was standard practice in the 1960s, Anne's baby was taken away from her and referred to by doctors simply as a physical entity.

'Damien'

Although her baby was born dead, Sarah, in contrast with Anne, was encouraged to wash, clothe and hold her 'son', to whisper into his ear so breathing social life into his body. Her child was named Damien. She could also hear him, even though *'it was not making any noise.'*[3] This was a silence she could hold in her memory as a part of who he was. And through this physical intimacy, whenever he was wheeled in to her 'in his small bed', little by little she discovered her child and who he was. At the same time, she came to realize that she was also saying goodbye to his body. However, the image she was forming of him, a 'real life' image viewed with her own eyes and not just in her head (as Anne did), was one she tried to keep forever. Even though his tangible body would soon disintegrate during cremation, Sarah's efforts to install him in her imagination through her interaction with his small lifeless body enabled her to fix him in her memory and keep him there alive. What is more, Damien's fleeting embodiment was caught on camera.

Unlike someone who dies in later life, however, Damien had no past. As Sarah explained, *'an adult leaves memories behind. You can talk about an adult who has died. You can say: he, or she, was like this, or like that.'* Yet her only living memories of Damien were of a child in the womb, evidenced by his ultrasound images. She added, *'The most I have from him was from the time he was still in my belly. Actually, he had been alive for nine months, but we've never seen him alive, at least not since his birth.'* Her son was thus 'born' before his 'birth' through embodied sensations and ultrasound images, making hers a *parental* experience. As Hockey and Draper argue, 'models of the life course need to accommodate the meanings of pre-birth and post-mortem materiality, and so incorporate a conceptualization of social identity as contested, relational and inevitably incomplete' (2005: 41).

Sarah's account indicates that stillborn babies are shaped and constantly reshaped in their parents' imagination, a process resourced by the relationship between actions, objects and spaces. For example, contemporary hospital practices mean that in parents' eyes their baby is now treated as if it were alive. Another participant, Patricia, whose baby was stillborn in 2004, explained, *'they took our baby immediately, just as they would do with another baby, to clean it a little, to dress it, and to lay it in my arms.'* Whenever Sarah and her boyfriend asked to see their baby, they were given his body. Unlike Anne's experience in the 1960s, their baby was not the hospital's possession. He was *their* child. For Sarah, the clothes she had herself bought him, though greatly oversized, were crucial: she said she *'couldn't care less: as long as he had [her] clothes on'*. Once wrapped in fabric she had chosen for Damien, his little body came to life, he became a social being. Furthermore, passing Damien from one pair of arms to another, from his grandma to his brother and to their best friends, transformed his body from some*thing* to some*one*. In this way, Damien came to exist through acts which connected him with people who were part of his parents' social identity (see Van Reusel, 1999: 87).

Damien was, however, simultaneously an impersonal 'it'. The temperature, colour and inertia of his body were empirical proof to Sarah that Damien was not biologically alive. In common with other parents' accounts of their child as a biological entity, she sometimes stopped talking about 'him' and used the more distant 'it' and 'that'. As Patricia explained, *'Obviously you do see that it is a dead child. It shows the signs of a corpse.'*

Tensions between Damien's (biological) absence and (social) presence also extended beyond the delivery room or hospital ward. This was echoed in Anne's account. Even after dissolution through cremation or burial, the absent child could remain socially present *within* things and places. Objects such as pictures or clothes, and locations including the grave and the nursery, could become extensions of their absence.

The cemetery

While cremation ended Damien's bodily existence for Sarah, when she visited his plot in the cemetery she perceived his absence as a *marked* absence. Both the place of the gravesite and the objects placed there not only referred to his death, but bore witness to his presence in the lives of other people, particularly of his parents: they helped him to remain socially alive (Mulkay, 1993: 30). Forty years earlier, Anne's child had a different destiny. It took 23 years for her husband to undertake a quest to locate their daughter's body. For all those years, Anne had suffered lugubrious imaginings: *'Maybe they had it preserved in formalin.'*

A number

For Anne, the shape and extent of her baby's life course were constituted through institutional rules and actions. Legally, the child was unable to bear a name; and church rules permitted neither a baptism, nor a funeral or a visible place in the churchyard. Without a social history, a past, the baby had no place in the public space of commemoration. For Anne, her child, 'that child' as she called it, had no marking, instead 'dissolving' into the row between the other graves. From 1966 until 1990 it lay there '*next to another grave*', '*not underneath a tombstone*', '*not below a name*', '*not under a grave*', but '*under a number*' visible only on the burial register. A TV film shown in 1990 about a couple who saw and buried their stillborn child led Anne and her husband to request a copy of the cemetery's burial plan. Although only a number on the plan, from that time onwards this document at least was in their hands; more than a bureaucratic document the identification of their daughter on the plan endowed it with social and emotional meaning (see Seremetakis' notion of 'sedimentation', 1994: 7).

Another interviewee, Paula, said she had never received an answer to the question of where her stillborn brother went in 1953: '*Where it was put, we don't know. At least, I don't. I never heard a word about it. Nothing.*' But she does '*know that her dad rolled it in a sheet and left by bike*'. Martha's baby also disappeared the moment it saw the light of day at home in 1951. The gravedigger knocked on the door almost immediately and for years Martha could still hear the sound of his clogs as he carried the little body behind their house and headed to the churchyard. There it was buried in unblessed soil, without a funeral or anyone present. Instead of resting in peace in a fixed and visible point, Martha felt her child would wander about forever like a ghost.

The empty space where Anne's daughter was buried served as an instructive metaphor for how she should construct the event of stillbirth; Anne was not meant to nurture the relationship with the baby, for instance, by tending the grave regularly or memorialising at its site (see Francis *et al.*, 2001: 228–229). Had there been a grave, this would have manifested her child's absence through its presence, an absence thus made explicit through material objects (Law, 2004: 84; Meyer and Woodthorpe, 2008).

The mushroom

By contrast, Sarah knew the location of her son's ashes and tended his grave. The ceramic mushroom she set up in place of a headstone was important to her, a way of ensuring that Damien did not '*get an ordinary grave*'. His 'different' grave thus materialised his different death. Leaving only minimal traces in his parent's world, a heavy marble gravestone for Damien would have been grotesque in Sarah's view. While not all Flanders' cemeteries offer such mushrooms to contain a child's ashes, stillborn children's graves usually look different: smaller, lighter, more playful, fragile and vivid – and showered with gifts (see also Francis *et al.*, 2001: 229; Garattini, 2007: 197).

Sarah also took comfort from knowing that Damien lay in a 'natural' environment, even though 'nature' had caused his death and scarred Sarah badly. 'Nature' was now, therefore, configured as a source of strength – and in the natural environment of the grave Damien had 'given' her, Sarah could find peace. For example, mostly she cycled alone to the cemetery, the mushroom quietly anchoring her son in the ground, rather than leaving him to roam freely, everywhere and nowhere like Martha's child. Through this configuration of objects and practices, a place for calm, control and reflection emerged, with the gravesite becoming a sacred place of transcendence. Over time, Sarah began to photograph it, storing identical – yet evolving – images of the mushroom in a small album: seasons changed and new objects entered these pictures (Figure 5.1).

Through the materially grounded practice of making an album of the mushroom, photographed time and again, but in a changing context of the evolving seasons, the stillborn baby was being given a life course.

Later Sarah brought her newborn second son to the cemetery and photographed the ensemble of pram and mushroom, so enacting her stillborn son's new identity as a brother. The resulting object – the photograph – then became a tangible memory of her sons' 'past experiences' together. This example contrasts markedly with Anne's experience, yet the historical shift in thinking and practice is only partial. Just as the body of Anne's baby disappeared into unmarked soil, so the precious album of photos was something Sarah, at least in my interpretation, perceived as irrational and shameful – at first she didn't want to show it to me, just as she didn't show it to her friends or family members – and kept it hidden in a cupboard.

Yet Damien's place within the lives of his parents and extended family did take a material form. When Sarah's mother bought wind chimes for her other grandchildren, Damien got some too. At Christmas and Easter when 'family' became a tangible reality, the mushroom was a powerful focus of activity where Damien was present as a social actor. The family members took him Christmas and Easter gifts. The 'new flowers'

Figure 5.1 Damien's grave in the summertime.

growing, dying back and sprouting nearby each year further served to locate his death within yearly cycles of growth, loss and return (see Layne, 2000; Layne, 2003: 116).

Marginalised from the taken-for-granted life course that is bounded by birth and death, Damien had a life course of his own, one that demonstrated the limits of a womb-to-tomb model of human life (Harris, 1987: 20; Hockey and Draper, 2005: 43). Challenging the narrative structures of how a life should be, things that referred to Damien, both before and after his death, confirmed his parents' desire for a biographical trajectory (Layne, 2003: 339).

For Sarah the mushroom was also associated with the image of a gnome, just as it was for Damien's godfather when he addressed Damien during his farewell ceremony: '*Wherever you are, you little gnome, we wish you all the best.*' For many other parents stars were the means by which they related to their stillborn child. Although at first sight gnomes and stars might seem radically different, their metaphorical use can be similar. Both images help parents to relate to their stillborn child in a material and spatial way. Just like the child, gnomes and stars are ambiguous. They are absent and present at the same time. They are close and far away. They are here and there. They are part of our time and space and they are not part of it. They are vivid, and they are small. In addition, however, gnomes are different from stars. Whereas stars are high in the sky, flirting with other worlds, gnomes are earthbound. Although stars are empirically described in physics, gnomes belong to the realm of fairytales and fantasy. While neither carries connotations of death, each, nonetheless, has properties that can be transposed metaphorically to the unimagined death of a child.

The home

While home has often been neglected in the cultural and historical study of loss and memory, for instance in the context of war (see Winter, 1995), domestic space and practice have been recognised in recent anthropological literature as a potentially important aspect of embodied experiences of death and grief (see Hallam and Hockey, 2001: 1).

'*The dead keep dwelling in the house,*' said Margret on the death of her husband with whom she had lived for 45 years. But when her baby died in 1950, that was different; in contrast to the death of her husband with whom she had lived for many years, the death of her baby did not '*shake the house*'. There were no worn-out shoes, no little notes, shopping lists or other lingering memories to stumble upon, once so trivial but after death newly visible and tangible (Hallam and Hockey, 2001: 12). With the death of a stillborn baby '*you have no reality of that child*', says Mary who lost her baby brother in 1959, or was it 1960 – she cannot remember. '*It has been. And it has not.*'

For Mary and other women such as Margret and Anne who experienced their loss in the 1960s and earlier, the unused clothes in the cupboard were material traces, not of the past, but of a future yet to arrive. Such objects gave a concrete and assignable form to the absent and ephemeral child. In the 1960s only tactile manifestations not stemming directly from the baby's body were possible – for instance, the cradle waiting in the living room for the baby to cry, sleep and awake in. By contrast, in Sarah's time, parents of a stillborn baby were given access to and control over extensions of the body itself – ash, hair and footprints, along with direct representations such as photographs, whether assembled on a 'shrine' in the corner of the room or kept in a box or album.

To illustrate this historical shift and how within the domestic space of Anne's time, the absence of the stillborn still materialized, despite the lack of memorial prints and pictures, I will use two objects from my data – a cradle and a doll.

The cradle

The cradle set up for Anne's baby was an object that had a social life, in that it manifested within a whole series of family life course transitions (Appadurai, 1986; Kopytoff, 1986). Anne prepared everything in the home to welcome her baby. *'Very symbolically'* she renovated the cradle in a way that also stood for or represented her: the cradle in which her husband's sister had slept years before was dressed with her wedding gown. She unpicked the fabric that had enclosed her body on her wedding day and used it to make the cot bedding. By doing this she symbolically sacrificed her old self for that of a mother.

While Anne was still at the hospital, following the baby's death, her husband Francis '*broke down the cradle*'. She told me, '*There was nothing left to see. He had silently cut off the ribbons, put all the pieces in a sack, and then removed the cradle. He was crying hard, he had told her, as he undid all the ribbons, one by one.*' Francis was well intentioned: he wanted to protect and care for Anne. And for him, moving on from the death, rather than holding on, was the 'proper' way to grieve. Dismantling the cradle could therefore be seen as an embodied performance that gave material form to the metaphor of 'breaking bonds', which underpinned the medicalised models of grief of that period (Walter, 1999; Bleyen, 2005: 189–204).

However, forgetting her baby was not an option for Anne. To this day the remains of the dress used as bedding are in a bag in the house. Anne says she '*should remove it one day, get rid of those things. But they are so hard to dispose of. But I cannot have it lying there forever. Then, my son will have to do it. I should tidy it up. Have it all gone. I'll keep it in my heart.*'

Like the empty space in the cemetery under which their baby's body was secretly buried, removing the cradle created an empty place in the house. Yet, for Anne this 'absence' materialised a total presence. She and Francis still saw the cradle in their thoughts; those thoughts would often evoke a baby sleeping or crying, moving its little hands and feet or being silent.

Thus, not only did the absence of the cradle embody the *loss* of the baby, it also made the child omnipresent for both its parents. Rather than 'too little', their baby's death created 'too much'; it took up the entire space of the home. Yet, while everything had painful associations with the baby, knowing that the cradle and the bag containing the remains of the wedding dress were still in the house comforted Anne. By sheltering these residual items, the presence of the bag softened the atmosphere produced by the unbearable absence of the baby-that-never-came-home. In the end, the bag became the shelter of a child who otherwise would have floated through the house, ghostlike and uncontrollable, lacking any material attachment (see Moyaert, 2007: 166–167).

In contrast, Helen, whose baby was stillborn in 1963, did not remove the cradle she had made for the baby. She changed it. By hanging a little cross on it, she confirmed that something profound had happened. She intervened in the object like God, or whoever or whatever had intervened in the baby's life. By marking the object, she brought a presence into the cradle that would otherwise have stayed empty. Helen added the sign of death to the sign of new life. Visitors to the house would notice how she had apparently 'accepted' the reality of her baby's death: how the cradle had become something akin to a grave for her baby. Eventually, she dismantled the cot and kept the little cross on her bedside table.

In 1990, Katy also left her stillborn child's cradle intact but shifted its position in her home, placing it prominently near the door to the living room. She told me, *'People almost tripped over the cradle which almost obliged them to talk about our child. Otherwise, they wouldn't have done so, because they were too afraid, whereas we wanted to talk about it so badly.'* Similarly, most of the more recently bereaved participants in my research – whether or not in response to self-help groups or literature – put a candle on the table with the baby's name on at family parties such as Christmas. In this way they retained the presence of their child and prompted guests to ask about their baby and stillbirth.

The doll

Alongside the cradle, objects such as toys, clothes and plants, though not derived from the baby's body, could subtly embody the baby's social presence. When Elisabeth's son was stillborn in 1960, her 4-year-old daughter was *'very sad because she had always hoped for a little brother or sister'*. Elisabeth tried to console her by giving her a toy: a new doll. It had to be a boy; that is, it was dressed like a boy. Whenever she played with it she was comforted. And when she went to bed she told her little doll: *'your brother is now in heaven, but you have to stay with me so I'm not alone.'* The doll obliquely 'replaced' the little brother, said Elisabeth, and became a point of reference for him and enabled embodied interaction. Being both absent and present, the dead baby was given entry to the perceptible world by the doll. Although the toy was treated as if alive, appearing to make her daughter's 'little brother' a tangible reality, the 'little brother' nonetheless remained intangible. Each time the mother and the daughter touched the doll, 'little brother' would slip through their fingers (see Moyaert, 2007: 29–30). Moreover, it was less a case of them expressing their emotions than the doll acquiring the agency to partly take over their role, as mourners, moderating their painful need to relate emotionally to 'little brother' (see Moyaert, 2007: 147–149).

Similarly, Sarah's photographs of Damien's body, dressed in his own clothes with his teddy, acquired agency with the result that the photographs retained her memory of her dead baby and enabled her partially to forget Damien and live a life without him. In other words, the photographs enabled her to do what she did not want to do and felt unable to do – since grief for Damien was the only emotion she felt capable of. Thus the photographs somehow absorbed her sadness, making her grief tolerable. Taking on her sadness, they taught her to live with the loss less painfully (see Moyaert, 2007: 151–152). Like the doll, the photographs meant that Damien did not have to become a ghost. Instead, he could, as Sarah told me, *'disappear and obey the rhythm of presence and absence'*.

Conclusion

Today in Flanders only six out of a thousand viable babies are stillborn, making such deaths an unusual event.[4] As such they provide a distinctive lens through which we can examine how, through historically specific practices, not only death but also life and a life course are performed via objects within particular spaces. Stillbirth therefore offers a useful opportunity to understand how 'the relationship between embodied individuals and the physical world around them' (Graves-Brown, 2000: 2) has changed in contemporary western societies.

In this chapter, I have focused on social actions within three interconnected spaces – the hospital, the cemetery and the home – in order to demonstrate how, through the enactment or embodiment of metaphoric images, death, grief and the stillborn baby became tactile and tangible realities. I have described how the concept of death came to exist in terms of the negotiation of spatial distance and intimacy, how mourning was materially performed either as 'letting go' or as 'continuing bonds' (see Klass *et al.*, 1996; Walter, 1999; Valentine, 2008) and how the dead baby became embodied as an immanent but 'other' being.

My core theoretical entrée into this material has been via work on metaphors. As I have argued, metaphors offer some potential for making order out of chaos; they can enable fragments of life to converge into 'things' that give a sense of wholeness. Thus, a metaphor such as 'time is money' underpins experiences of time as something we can save, win, lose and invest. Although we are hardly conscious of this, without metaphors we cannot communicate and make sense of life (Lakoff and Johnson, 1980: 453–457). Indeed, metaphors are omnipresent in daily life, not only in our thoughts and conversations but also in our embodied actions. Metaphors are therefore neither merely subjective nor objective. Rather, they are inter-subjective (see Hockey, 1990). Moreover, being performed in social contexts they become vulnerable/open to change. This chapter has drawn on these theoretical perspectives to make sense of two key case studies within my data.

In the 1960s, when Anne's baby died, hospital treatment rendered such babies 'dead' and object-like through sets of practices which embodied particular metaphors of grief. For example, in order that parents could 'let go' more easily, the baby's absence rather than presence was promoted. In the hospital, the baby's body was hidden immediately. Parents were not told where and how their baby had been disposed of in the cemetery and had no particular place to visit and tend. At home, Anne's child's father dismantled the cradle. Paradoxically, however, such absences could serve to evoke a powerful presence for the stillborn child.

Forty years later, Sarah was encouraged to interact with her baby's body as a subject – which was both dead and alive – in order to continue and create bonds. Indeed, as stated by Hallam and Hockey,

> Western practices that have developed around the prenatal deaths of infants, such as photographing and measuring the body, giving it a name, a death certificate and a grave, are recent cultural innovations. Prior to the 1970s, foetal remains were disposed of with no

> religious or institutional attempt to provide them with a social body and therefore a memorable identity.
>
> (2001: 4)

Such items, objects and practices materialised a new set of metaphors which framed stillbirth as the death of a person rather than the outcome of an unproductive pregnancy. Now, stillborn babies may be represented as 'stars' or 'gnomes' – embodied metaphors that materialise at the gravesite, evoking the child's ambiguous status as both presence and absence. Furthermore, as I have argued elsewhere (Bleyen, 2009b) through these embodied interactions with objects and spaces, the concept of 'death' has shifted its boundaries and conceptualizations.

Notes

1. I deliberately do not define stillbirth in specific medical terms with reference to gestation period or birth weight, since these categories in themselves are subject to change. Instead, I define stillbirth as the loss at birth of a baby which was viable and considered to have lived a biological life only within the womb. For a discussion of medical terms such as miscarriage, stillbirth and perinatal mortality, see Layne (2003: 9–11).
2. In contrast to countries such as the United Kingdom, the Flemish marriage certificate lists the couple's children.
3. I will discuss the disjuncture between 'him/her' and 'it' later on.
4. Perinatale activiteiten in Vlaanderen 2004 (http://www.zorg-en-gezondheid.be).

6
Distributed Personhood and the Transformation of Agency: An Anthropological Perspective on Inquests

Susanne Langer

This chapter focuses on the ways in which culturally specific ideas about persons and objects influence the expression of agency in the inquest process. In England and Wales, under certain circumstances, deaths have to be investigated by a coroner. In 2008, 234,800 deaths – or 46 per cent of all deaths in this period – were reported to coroners (Ministry of Justice, 2009: 3). However, at 31,000 – or one in eight – the number of inquests opened was significantly smaller (Ministry of Justice, 2009: 5). A post-mortem had been performed in 92 per cent of these cases. While there has been a general downward trend of post-mortems conducted, the overall number of deaths reported to coroners has been rising steadily in recent decades (Ministry of Justice, 2009: 5).

Deaths reported to the coroner are those that have played havoc with deeply held assumptions about what constitutes a 'good' way of dying (Bloch and Parry, 1982). As the examples below demonstrate, such deaths are sudden, or unexplained, or occur at the wrong time, place or for the wrong reasons. The ensuing investigation generates a considerable number of documents, including expert reports and witness statements. It culminates with a verdict reached at a hearing at the coroner's court. Here experts and witnesses, including those who have been bereaved, come together to establish who the deceased person was and how they came by their death. Inquests do not address questions of guilt and culpability, yet such concerns tacitly inform much of the proceedings. Given this remit, I draw on Gell's (1998) ideas to argue that inquests become the site of a collective transformation of agency, which also implies a transformation of the kind of persons involved in the inquest. In so doing, inquests serve to – albeit temporarily and partially – reassert culturally specific notions about persons and about death.

Empirically, this chapter draws on documentary research into coroners' files of inquests concluding in a verdict of suicide, as well as participant observation at a number of hearings. In England and Wales, coroners and inquests have mainly been studied by Prior (1989) and Howarth (1997, 2001b). While drawing on both authors, my focus here is neither on the centrality of pathology to the inquest, nor on the implications of the inquest's silence on issues of culpability. Instead, using anthropological theory, I focus on the ways in which the issues of agency, objects and personhood are played out in the inquest process.

Specifically, Gell's (1998) work on distributed personhood and agency and Mauss's (1987) arguments on the category of the person are relevant here. Gell proposes a model of personhood that is obscured by the western concept of the autonomous individual. While for Gell, agency is crucial to being a person, this capacity is not situated in the agent itself, but has to be constantly generated through a network of relationships. Even 'things' can be agents, he suggests, but in contrast to human agency, the agency of objects excludes the capacity to choose and the moral implications choice brings with it. He refers to this distinction as primary and secondary agency. I extend Gell's point about agency by arguing that under certain circumstances, such as an inquest, the primary agency of human beings can become a form of secondary agency, akin to that of objects. This implies that within the network of relationships that constitutes the coroner's court persons, as individuals, turn themselves into objects, in a way reminiscent of what Mauss has called personas; that is, they act as representatives of a particular social role, with consequences for their ability to be held responsible for their acts. Such a process is always partial, and aspects of individuals and personas are intertwined in the persons acting during an inquest. It is the oscillation between individual and persona that lends the inquest some of its potency.

Successful inquests, I suggest here, are about the transformation of primary agency into secondary agency, because commonly they present the death as the inescapable consequence of the intersection of a multitude of events. By transforming the death into an incident that could not have been prevented, it becomes possible to repair and reaffirm culturally dominant assumptions about the 'facts of life'. I will trace this conversion process through some of the documents of the inquest, as well as through the public hearing. First, I will provide further information about the research and introduce the coronial system at greater length.

Research methods

This chapter's empirical data are derived from two studies. The first was a demonstrator project of the Cardiff University node of the ESRC National Centre for Research Methods. The other was a small-scale investigation of 17 inquest hearings in three coroner's courts.

The demonstrator project combined documentary research into inquest files ending in a verdict of suicide, with a small number of interviews with family members and friends of the deceased person, and in some cases media reports, in order to shed new light on suicidal lives (Shiner *et al.*, 2009). For this chapter, I will focus on the 100 files from the coroner's office of a medium-sized British city that constituted the main body of data for this project. Access to the files had been granted by the coroner, and word-processed copies of key documents, such as witness statements, reports by medical professionals and police officers, and suicide notes were produced on site, as well as notes recording the structure of typical files.[1]

The second source of data utilised in this chapter derives from 17 hearings I attended at three coroner's courts, serving either medium-sized cities or larger conurbations, with four different coroners presiding. Inquest hearings are open to the public and – in theory – anyone can attend. In practice, few people other than the interested persons do so.[2] Before the hearing began, I would introduce myself to the coroner's clerk or court usher and then settle towards the back of the courtroom. After the hearing, I would introduce myself the bereaved family and express my condolences if this seemed appropriate.

The role of the coroner

Coroners are independent judicial officers whose duty to investigate deaths is linked to both how and where a death occurred. Hence they inquire into deaths that have been sudden, violent or unnatural, or where the cause of death was unknown. Yet they also investigate deaths that occurred in instances where a person's autonomy was compromised, such as in custody or if they have been detained under the Mental Health Act (2007).

The official discourse pertaining to inquests defines them as non-adversarial inquiries where no one is on trial and which are not concerned with questions of culpability and blame but only with establishing the facts surrounding the death: who the deceased person was, and how, when and where he or she died. Where a criminal offence has caused the death, this will be dealt with in a separate trial. Once these questions have been answered to the satisfaction of the coroner, he or she will reach a conclusion regarding the cause, or rather the category, of the death. Over the last decade the most common ones were accident or misadventure, natural causes and suicide (Ministry of Justice, 2009: 7).

'Good' deaths – 'bad' deaths

The deaths coroners deal with are largely 'bad' deaths, because they contravene expectations about where, when and how a person should die (Kellehear, 2007). Bloch and Parry, for instance, declare suicide – especially in Christian cultures – to be 'the supreme example of a "bad" death' because unlike its counterpart, 'good death', it lacks the potential for rebirth and regeneration (Bloch and Parry, 1982: 16). These authors also identify a certain degree of control over the arbitrariness of biological life and a reaffirmation of the continuation of the social order to be important elements of a 'good' death.

However, other research calls some of Bloch and Parry's analysis into question. For instance, the Vaquieros, Catholic pastoralists of Northern Spain, consider death by suicide to be preferable to a long, lingering and painful 'bad' death (Catedra, 1992); and in her study of representations of the 'good' death in England, Bradbury found no evidence of deaths being classified according to their regenerative potential (Bradbury, 1996). In support of Bloch and Parry though, Bradbury's research shows that control is indeed an important factor that distinguishes 'good' deaths from 'bad'. Control, she suggests, refers to the ability to influence the time and place of one's death and the people present. A 'good' death is also largely painless. In material terms, a 'good' death happens in old age, in a peaceful place, with loved ones present and with any pain well managed (Bradbury, 1996: 94).

As noted, these are conditions that are rarely met by the deaths coroners typically deal with. Some of the above conditions, such as suddenness or unknown causes, can also apply to deaths in general hospitals and the coroner will investigate these deaths, too. Hospital deaths can be particularly disturbing because they challenge the promise and power of modern medicine and question the belief that hospitals are places where people get better, rather than die (Porter, 1999). In this sense, the deaths coroners deal with are deaths that *should* not have happened, or at least not under these particular circumstances, because they contravene normative expectations about death.

Thus 'bad' deaths play havoc with cultural assumptions about death and dying and in an argument that recalls Hertz' analysis of funeral rites, Howarth proposes that inquests are a way of reordering and controlling apparently random deaths (Hertz, [1907] 1960; Howarth, 2007). I agree with her interpretation of inquests as acts of ordering, but rather than Hertz's ([1907] 1960) concern with the relationship between society and its individual members, my approach draws attention to the diversity of agents, the transformation of agency and the distributed nature of personhood as crucial factors in such a reconstitution of the taken-for-granted 'facts' of living and dying that is the inquest.

During an inquest, I suggest, those affected by the death will present their own agency as compromised by other forces. While doing so reduces their ability to be agents, it also diminishes their potential to become the target of accusations and blame. In a sense, the interested parties in an inquest turn themselves into 'objects' or what Mauss would have called 'personas', that is, they become representations of a particular category of person, such as a consultant, a police officer, or a grieving relative (1987). How convincing such a transformation is depends on the assumptions inherent in the institution of the inquest about the degree of 'objectivity' or 'subjectivity' that different witnesses bring to the inquiry. For instance, while the accounts of experts are considered to be 'objective', those of lay witnesses, such as friends and family members of the deceased person have to be translated into 'objective' statements by the coroner (Howarth, 1997). How this is possible will become clearer, once the concept of distributed personhood has been explained in more detail.

Distributed personhood

This concept has its origin in the anthropology of Melanesia and is closely associated with the work of Strathern (1988) and Wagner (1991). In their view, personhood is constantly created through complex networks of relationships rather than being inherent in the individual. Not only does this insight turn key intellectual concepts such as 'individual' and 'society' on their head, it also shifts the focus of sociological and anthropological inquiries from the relationship *between* individuals and society, to the practices that make human beings *emerge* as particular kinds of persons. A crucial role within such practices is played by agency.

For Gell (1998), on whose work I am primarily drawing here, agency is not simply an expression of physical causality, it also encompasses intentionality. Furthermore, agency is always social and although ultimately located in humans – because only they can have intentions – agency is not restricted to them. In the context of human social interaction, objects can have agency attributed to them and human agency can be exercised through these objects. Like humans, objects can make things happen, but unlike humans, no alternative decisions are possible for them. This is why, in this view, the agency of objects will always remain what Gell refers to as 'secondary' agency in contrast to the 'primary' agency of humans. As noted, in the inquest proceedings individuals themselves may seek to take on the status of personas, so transforming their primary agency into secondary agency.

Gell illustrates this distinction with an example from the minefields of Pol Pot's Cambodia (Gell, 1998). The mines' terrible effect is not a consequence of their intentions, but of their compulsion to explode once touched. This makes the mines' agency different from the agency of the soldiers who laid them and exercised their agency as soldiers through them (Gell, 1998). Whether a soldier accepts their primary agency in causing death is not a foregone conclusion, as the example of the infamous Nuremberg Defence has shown. Nor, is such a defence evidently untrue because the military and other institutions are geared towards limiting the extent of primary agency.

While the example of the mines and the soldiers may be extreme, it helps explain what happens at inquests. Not only do inquests offer an illustration of the social framing of agency, they also highlight the role objects, such as documents or the layout of a room, can play in the transformation of primary into secondary agency. Unlike war crimes tribunals, inquests do not deal with the moral question of guilt, but only with the factual circumstances of how someone has died. The issue of guilt is never directly addressed in an inquest and during hearings coroners go to great length to ensure that the interested persons and the audience are clear on this point. In instances where the death was caused by a criminal offence, this matter will be dealt with in a separate trial. By situating the question of guilt outside the scope of the inquiry, inquests mute the role of primary agency and elevate the impact of secondary agency, whether that of personas or objects. Thus, the transformation of agency – the idea that whatever happened could not be helped – is already built into the coronial process.

Persons, documents and agency in the inquest file

In their inquiries, coroners draw on a wide range of information, much of it related to medicine. The primary document that allows the coroner to establish the medical causes of death is the pathologist's report of the post-mortem of the deceased person's body. If necessary, this will be supplemented by further analysis, for instance, of histopathology or toxicology. The deceased person's GP, or if they had a history of mental illness, their psychiatrist may also provide a report. Yet, by searching for an answer to the question of 'how the deceased came by their death', coroners are not only looking for medical phenomena but also inquiring into the wider circumstances of a death. This means family members and friends may be interviewed, as well as those who were present at the scene of death, either as accidental observers or in a professional capacity, such as police officers or members of the emergency services. While the inquest file provides the basis for the coroner's decision, coroners value the hearing as an opportunity to clarify matters arising from the documents of the file and to provide new insights into the circumstances of the death. On the basis of all the evidence presented, the coroner will reach a conclusion as to whether the death was natural, accidental, due to suicide or a result of other causes.[3]

Before turning to the significance of these court hearings in the inquest process, I will discuss the material aspects of some of the documents that constitute the inquest file. Drawing largely on data generated during the Sociological Autopsy study, I will concentrate on those documents that are most significant in an inquest: pathologists' post-mortems, reports by physicians and witness statements. A closer look at these documents is instructive not only because of their content, but also because they provide insight into the materiality of ordering 'bad' deaths, and into the process of creating different categories of persons and of transforming their agency.

Post-mortem reports

The post-mortem reports in the files replicated the pathologist's progress in his or her examination of the corpse and left little room for individual expression.[4] The reports began with a brief summary of the circumstances under which the death occurred, including the estimated time of death, as well as identifying details of the deceased person.[5] Pathologists then proceeded to note external features before entering the body. External examination would establish such things as the condition (e.g. refrigerated), gender and state of nourishment of the body (e.g. well nourished), as well as noting outward signs of injuries, fresh and old scars, evidence of medical interventions (e.g. catheters) or identifying features such as tattoos. Then, the pathologist would direct his or her attention inwards, investigating the body's systems, including its organs. At each step, he or she would note evidence of injuries or any other signs of diseases or abnormalities, such as unusually large organs: a common side-effect of the deceased person having been bed-bound. Other signs of pathology would also be reported, such as clogged or enlarged arteries, the presence of fatty deposits in the liver that are an indication of prolonged alcohol abuse, or the existence of granular cells in the kidneys that point to a history of hypertension. Signs of trauma, such as haematomas of the brain, or broken bones were also reported.

As documents, post-mortem reports are commonly presented in a standardised format. Berg and Bowker (1997) suggest that such a limitation of individual expression is a sign that the medical specialist in question occupies a comparatively low place in the professional hierarchy. However, in relation to inquests, it is precisely the relative absence of individual agency that lends pathologists' reports their power, because it suggests that their actions have not been influenced by any subjective interests, but instead have been guided by objective concerns. In their reports objective inquiry – or fact-based investigation – was exercised through the *person* of the pathologist and the facts were speaking through them. This impression was not only supported by the documents but also supported by the way in which pathology delineates its subject of inquiry.

Pathologists' approach to the body as a virtually autonomous system creates a narrative of causation in which human agency remains only diffusely discernable. Hence, even where falls or surgical procedures were preconditions of the death, in the context of the post-mortem they are only relevant in so far as the act itself – but not the persons involved in it or why it had occurred – had precipitated death.[6] Thus by conceptualising death and disease as located within the human body and by downplaying the evidence of individual agency in their reports, pathologists made an important contribution to presenting inquests as fact-finding inquiries that privileged secondary agency.

Statements by medical professionals involved in the treatment of the deceased person

Because the coroner's remit is not only to establish the medical causes of death but also to investigate the circumstances leading up to it, those medical professionals involved in the deceased person's care can be requested to give evidence. This includes health care staff attending to the deceased person in the period immediately preceding their death, as well as GPs and psychiatrists who played a role in his or her longer term care. As most people nowadays die in hospitals (see Chapter 2), the chance to find themselves called to give evidence at an inquest is especially high for hospital-based staff (Ahmad and O'Mahony, 2005).

While sharing with their colleagues in pathology the discourse of clinical medicine and appearing at the inquest in a professional capacity, the purpose and position of health professionals involved in the patient's care is rather different. Their role is not to establish the causes of death, but to shed light on the circumstances leading up to it. As explained previously, the deaths coroners deal with contravene expectations about the potency of medicine and the circumstances of death. Furthermore, these health care professionals first encountered the deceased person as a patient – rather than as a corpse – which means the question whether they were culpable of the patient's death was not as far removed from them as it was for the pathologist. Although inquests do not openly address the issue of guilt, it remains a latent concern, especially if the death followed from a medical procedure (Davis *et al.*, 2002). Thus, in inquests, medical professionals found themselves called to account for their profession, their institution, and also implicitly as individuals.

To an extent, this ambiguous position was already visible in their reports and statements. Unlike post-mortem reports, which resembled bureaucratic forms, the statements of medical professionals were narratives. If the appearance and structure of statements by medical professionals offered them greater individuality, their internal structure was similar to the patient records described by Berg and Bowker (1997). At the same time, these statements were not identical to the patients' records, but had been specifically created after the death and at the coroner's request.[7]

In addition, the treating health professionals' reports varied in size, depending on how long the physician had known the patient and the extent of the treatment the patient had received. Some simply confirmed that the deceased person had been registered with the GP, but that little else was known about them. Others provided detailed case histories documenting symptoms and interventions, sometimes over a period of years. They had in common a focus on the empirical and clinical. Although, such an emphasis on facts was encouraged by both medical practice and the inquest process, they also had the potential to appear cold and unfeeling, especially if read out during a hearing. Hence, the statements reflected the ambiguous position in which health professionals concerned with the treatment of the deceased person found themselves.

In the inquest, doctors and nurses who were involved in the deceased person's care appeared as both individual persons and professional personas. However, because they encountered the deceased person in the pursuit of their job, and because it was in this capacity that their insights were relevant to the inquest, their professional persona took precedence. Further, coupled with the inquest's focus on 'facts', rather than morality, both enhanced medical professionals' ability to transform their primary agency into secondary agency.

Other witness statements

The category of witnesses affected by the inquest in a non-professional capacity was diverse. It encompassed those who observed the death occurring, or who discovered the deceased person but who otherwise had no connection to them, as well as those who had known the deceased person intimately.

Witness statements were shaped by the conventions of police statements and concentrated on the immediately tangible and observable. This focus on facts, together with being co-productions between statement takers and statement givers, lent these documents a recognisable style and meant statements differed in content rather than in appearance. They opened by giving the name, age and contact details of the witness, as well as the relationship in which they stood to the deceased person, and then proceeded with a narrative that had been condensed and chronologically ordered but preserved some of the witness's own voice. While pathologists treated the deceased person's body as a particular manifestation of regularity and sameness – the evidence of haematomas, heart attacks and cancers looks recognisably the same in any body – witnesses were concerned with incomparable difference. This was because on the one hand, they had encountered the deceased person as a unique individual, and on the other hand, because they tended to have relatively little experience of comparing deaths.

The statements of lay witnesses were especially important in cases where suicide was a possibility, because they helped to establish whether the death had been intentional. For instance, these witnesses offered accounts of how the death had occurred or they testified to the degree of conscientiousness the deceased person had customarily shown with regards to their medication. Finally, they could provide information about the deceased person's mood shortly before they died and about any remarkable events that may have triggered a decision by the deceased person to take his or her life. Witnesses would describe this information in lay terminology, remarking on how the deceased person appeared distracted, how they had been feeling low, how they had muttered about not wanting to be alive anymore or how they had tried to give away their favourite possessions. For other witnesses though, there had been no warning signs, either because the deceased person did not reveal their plans, or the bereaved person was not alarmed, or the death was unintentional, as in the case of accidental overdoses.

The statements of witnesses who were close to the deceased person are especially interesting, not only because the official investigation could present them with rather uncomfortable questions.[8] Although the coroners were careful to ensure that the inquiry did not turn into a trial, the inquest's central question of 'how the deceased came by their death' always also involved the witness. In the case of suicide, as well as in circumstances where the death had been unexpected or where the cause of death was unclear, witnesses would stress their relationship with the deceased, while simultaneously playing down their involvement in the death, instead asking whether the deceased person, the authorities, the doctors, any medication or physical or mental illness might be the leading agents of death.

Finally, the witness statements were diverse in terms of who had made them. They were also often rich and ambiguous in terms of their content. They fulfilled a vital function in the inquest, but by asking questions about agency and causation, the witness statements also raised the spectre of culpability and blame. Hence witnesses tried to diminish their ability to influence the course of events and actions that preceded the death.

Persons, objects and agency in the coroner's court

Inquests are not, however, decided on the basis of documentary evidence alone, and often culminate in a hearing at which the coroner reaches his or her verdict. At this hearing witnesses and experts will be called to testify and to respond to questions by the coroner and by the interested persons. Hearings acknowledge the personal dimension of the death, but also continue with the process of transforming primary into secondary agency, by privileging certain accounts, in particular the narrative of the death created by the coroner. They also help to restore assumptions about humans' potential to control their environment. Furthermore, the hearing demonstrated that the death had been properly and thoroughly investigated and thereby validated the witnesses' accounts with their emphasis on that person's restricted ability to prevent the death. Finally, the hearing offered the possibility of aligning the death in question with notions of the 'good' death. As with the inquest documents, the material world played an important role in this process.

Hearings take place in coroners' courts. These are either spaces that are for the exclusive use of the coroner, or they are temporarily used by the coroner for the public hearing, but they have in common that they create a formal environment for the inquest. The template for the layout at hearings was court rooms, with the coroner sitting on a raised bench beneath the Queen's coat of arms facing the audience. His or her officers or the court clerk would take their seat below the coroner's bench. In a typical layout, the witness stand will be to the right of the coroner's bench. The rest of the space will be taken up by seats for interested persons and any other attendees. The first row is usually reserved for the deceased person's next-of-kin and for any lawyers acting on behalf of interested persons. The remaining seats will be occupied by further witnesses, friends and family; the press, if present, tend to sit to the side or the back of the court room. The layout of the court room thus serves to emphasise the power and independence of the coroner and their role as a representative of the state.

Both the setting of the court room and their procedures materially restore order to a death deemed chaotic. The audience is asked to rise when the coroner enters, witnesses are called and sworn in, and prescribed formulas and standardised expressions are invoked. There is also a clear hierarchy in terms of who is allowed to put questions to the experts and witnesses. The coroner will take the lead and guide the witness or expert through their statement, asking them to elaborate or explain in the process. Although other interested persons are only allowed to pose their questions once the coroner has finished questioning the witness, coroners will go out of their way to make it clear especially to the bereaved family that they have a right to inquire. Having said this, the coroner may provide guidance as to which questions are appropriate, in order to prevent the inquest from turning into a trial. Their ability to manage the interested persons ensures that the inquest conforms to its prescribed format, while simultaneously emphasising the limited degree of agency that can be exercised by the witnesses.

The coroner's role in transforming the death into an event that could not have been avoided is central to the inquest.[9] They were the ones who at the end of the hearing would merge the different accounts into a coherent narrative of the death. They were also able to shape the proceedings, for instance with the questions they pose. They can privilege certain accounts, for example, by emphasising how helpful they found them and by explicitly thanking the witness in question. This applies particularly to the pathologist's post-mortem report because his or her findings may not agree with the bereaved family's explanation of what had caused the death. And finally, coroners will pronounce the verdict and reach a conclusion. Not only does the inquest conclude with the hearing, but the transformation of agency that is central to the inquest is encapsulated in the persona of the coroner. The key elements of the inquest, such as the intertwining of the persons and objects, persons and personas, can be seen in the vignette below.

Mr Kelly, an older man, was in hospital for cancer treatment. While recovering from the operation his tracheotomy had become dislodged. Despite the immediate intervention of hospital staff, Mr Kelly slipped into a coma from which he did not emerge. Mr Kelly had originally been reluctant to undergo cancer treatment, but had changed his mind on the insistence of a close friend. It was the same friend who eventually gave permission to switch off Mr Kelly's life support and who acted as his next-of-kin at the hearing.

The key witness in the hearing had been the consultant who led efforts to revive Mr Kelly. The consultant was guided by the coroner through his statement and asked to describe what he had done and explain why. In his summing up, the coroner reiterated that the tracheotomy becoming dislodged was a recognised complication, stressed that a 'culture of responsibility' had been in evidence at the hospital, and that emergency resuscitation efforts had been appropriate and sufficient. He concluded that 'on balance, it was in [Mr Kelly's] best interest to have the surgery' and gave a verdict of 'died from complications arising out of necessary, complicated surgery.' This, he explained, was a verdict he had devised himself and used regularly for instances where medical interventions had been conducted correctly, and had nonetheless ended in the death of the patient. In proceeding through all the above steps, the coroner established that the death was nobody's fault, including the friend who had advised Mr Kelly to have the operation in the first place. The coroner performed a thorough investigation of Mr Kelly's death by taking his time to interview witnesses and establishing through his questions that the treatment and care that Mr Kelly had received could not be faulted. In addition, he also addressed the question of guilt in his verdict, by showing that the operation had been in Mr Kelly's best interests. Hence, it was not individual actions or intentions that were at fault, but circumstances over which the human agents involved had no control. In so doing, the coroner reasserted cultural assumptions about death, because he located its causes outside the social realm and labelled it as the exception that confirmed the 'good' death as the rule. This happened, for example, by stressing that everything had been done to prevent the death and that the deceased person would have died quickly and without having to suffer.

Conclusion

In this chapter my aim has been to introduce inquests as sites where sudden deaths are re-ordered. Using the concept of distributed personhood has been especially appropriate in this context, because the theory pays particular attention to the ways in which different kinds of agents, such as humans and 'things', become entangled in the creation of different kinds of persons. In addition distributed personhood allowed me to draw out some of the collective effort that is necessary for the conversion of agency that lies at the heart of the inquest. If successful the death will have either been no one's fault, or it will have been brought about as a result of an agent having no choice over their actions. This transformation of sudden death into deaths that are unusual, but do not invalidate deeply held beliefs in the 'facts of life', requires considerable resources in terms of people, skills, time and material. It can also remain incomplete, because while the intention is to minimise one's own role in the death, this does not necessarily apply to other agents' roles. Hence, in attempting to de-emphasise their own agency, interested persons may highlight that of another party. Nor does the inquest process constitute a level playing field. With their emphasis on the medical causes and their construction as a non-adversarial fact-based inquiry into the death, inquests privilege some accounts over others. Nevertheless, it is important to recognise inquests as collective productions that involve the active engagement of different kinds of persons and of objects.

Notes

1. We were allowed neither to take the files out of the coroner's office nor to make mechanical copies of the documents within them.
2. Only 'properly interested persons' are allowed to examine witnesses and have a right to inspect the files. Rule 20 (Coroners Rules 1984) defines them as next-of-kin of the deceased person, but also as those benefiting from any insurance policy of the deceased, insurers, anyone whose actions or omissions may have contributed to the death and their representatives, trade union representatives

in cases of suspected death by industrial disease, inspectors and the chief of police, and 'any other person, who in the opinion of the coroner, is a properly interested person' (Dorries, 2004: 145). Researchers are not legally considered as properly interested persons, but in our case, the coroner decided that the study's contribution to suicide prevention was sufficiently important to grant us access to the files.

3. In addition, inquest files also can contain photographs of the scene of death, suicide notes and some of the correspondence relating to the case. The files may also contain the original hand-written versions of forms and statements. For a comprehensive discussion of the contents of the files of inquests ending with a verdict of suicide, see Fincham *et al.* (2008).
4. The Coroners Rules 1984 even provide a pro-forma for a post-mortem examination report (Dorries, 2004, Appendix 5: 345).
5. For a detailed description of a post-mortem, see Stefan Timmermans' study of medical examiners in the United States (Timmermans, 2006: 49–63).
6. Causes of death are reported as Ia, Ib, Ic and II, where Ia is the most immediate cause of death, and Ib and Ic are conditions that led to the direct cause of death. II refers to significant underlying conditions that may have contributed to the death, but did not immediately cause it. To illustrate, the causes of death in an elderly men were recorded as Ia Fall, II Osteoporosis. The osteoporosis did not kill him, but the brittle bones in his neck broke, rather than cushioning his fall as would have been the case in someone without the condition.
7. If a patient died in hospital and one of the hospital trust's employees had been asked to provide a statement and appear in court, they would be accompanied by a member of the trust's legal team who would have advised them throughout the inquiry.
8. In the inquest, the witness that comes to represent the deceased person in their bodily absence is not necessarily a blood relative or an (ex-)spouse. For instance, in the hearings I attended, this role was taken by a former gay partner, rather than the deceased person's father, or by a male friend, even though the deceased person's ex-wife was present, too.
9. A qualifier is required here: coroners have the right to make recommendations to the appropriate authorities in order to prevent further deaths and they will exercise it. Although such recommendations have no legal power, they can have an impact, especially if widely reported in the media.

7

Behind Closed Doors? Corpses and Mourners in English and American Funeral Premises

Sheila Harper

This chapter can be located within a now-established tradition of research focusing on funeral rituals in different western societies, including England (Naylor, 1989; Howarth, 1996; Gore, 2007), the United States (Habenstein, 1954; Pine, 1975), Canada (Emke, 2002), New Zealand (Schäfer, 2007) and Sweden (Bremborg, 2006). This research typically considers the funeral industry, with funeral premises theorised as sites where funeral professionals establish and maintain their authority over the dead body. There has however been little research addressing the way mourners experience recently dead bodies in these settings. This chapter therefore focuses specifically on how mourners understand and interact with the dead body within the physical space of the funeral premises during the liminal period between the moments of death and the final disposal. I argue that the spaces in which mortuary rituals occur both shape and reflect viewing practices.

The arguments made in this chapter are based on findings from a cross-national research project designed to compare the viewing practices of two different western countries. Utilising ethnographic data generated from a funeral home in the United States (in this chapter referred to using the pseudonym, 'Blake's Funeral Home') and a funeral premises in England (here called 'Durnford Funeral Directors'), the chapter considers how the identity of the deceased person is (re)constructed during the liminal period. Significant differences emerge between the two sites in terms of how the ritual process of viewing the body, conducting the funeral, transporting the body to its site of disposal and, finally, committing it to either the earth or the cremator serves to make a very different contribution to the re-constitution of the social identities of the person who has died and those bereaved of a relative or friend.

This chapter will discuss how the liminal period unfolds by describing the various stages of ritual in each setting, so explicating how the deceased's identity is either obscured or revealed within the spatialised interactions that make up that ritual process.

The corpse and the self

Central to this chapter is the dead body. The dead body has been theorised as many things: an unwelcome reminder of decay (Featherstone, 1991), a site of information and contested interests (Prior, 1989), a failure of the body project (Shilling, 2003), a symbol of pollution (Douglas, 1966), a disruption of order (Douglas, 1966; Kristeva, 1982), a symbol of self (Synnott, 1992), the termination of self (Giddens, 1991; Shilling, 2003) and a presence that manifests an absence (Hallam and Hockey, 2001). More abstractly, Howarth (2001a: 120) identifies the dead body as signifying 'the loss of self and the loss of individuality – the material reality of death. As such, the dead body, once a symbol of natural order, now has a debstabilizing [*sic*] effect on social order.' Arguably because dead bodies are understood as polluting (Hallam *et al.*, 1999) and a symbol of disorder, dysfunction and danger, mourners in 'contemporary western societies' typically elect to award custody of the corpse to professionals. Howarth concludes by stating that

> (D)espite a cultural perception of the corpse as polluting, bereaved people continue to be emotionally attached to the body of their loved one. It is for this reason that death educators and counsellors recommend viewing of the dead body. This is thought to have a cathartic effect on the bereaved.
>
> (2001a: 121)

However, few theorists acknowledge that the dead body is not a uniform entity but one that might represent different things to different people at different times – or indeed different things simultaneously. Prior (1989) notes that the dead body is a site of information for the State (illustrated in the description of the coronial system in Chapter 6) while at the same time being a site of personhood for mourners. Hallam *et al.* theorise the dead body in terms of the negotiation of meaning, problematising 'the event or process we call "death" and the thing we call a "dead body"' (1999: 64). 'Death' cannot be identified as a 'straightforward moment or event which coincides with an obvious set of organic changes within the living body' (1999: 66); consequently, the dead body only comes into being as 'an outcome of the social relations through which the categorisation of death is negotiated' (1999: 69). This categorisation can happen on the part of anyone involved, including the medical professionals, the funeral professionals and mourners.

Another way of theorising the dead body comes from Williams (1990), who contrasts conflicting attitudes about the post-mortem self, the soul and the corpse, to examine the dichotomy in mourners' perceptions of the dead body as both something sacred and something that must be disposed of in order to 'minimise problems for the living'. Others are in agreement, suggesting that mourners differentiate between the unprepared 'polluted' corpse and the prepared 'sanitised' body (Howarth, 1996, 2001a; Hallam *et al.*, 1999; Gore, 2007). However, any fixed division between the sacred dead body and the profane corpse may be too simplistic, as it has also been argued that ritual processes can alter the meaning of dead body over time (Davies, 2002).

These authors, in their deconstruction of meaning surrounding the dead body and how it can change over time, forge the path that this chapter follows. However, an examination of the dead body and the setting in which it resides cannot be separate from theoretical discussion about the self and identity. According to Synnott (1992), the self is socially embedded. Linked to the body but also separate from it, the self arises from social experience (Mead, 1934: 140); it is not a concrete entity, but a 'process of behaviour' (Hewitt, 1979: 14). Of course, such person-centred definitions favour the living self. In their discussion of personhood, Hallam and Hockey suggest that it is through 'social engagement and networks of socio-cultural relations that persons and their identities are produced' (2001: 42). Such a conception of self removes it further from the living self as constructor, acknowledging social context as an important site of construction, and it is within such social contexts that the self and identity can be understood as interconnected. Jenkins (2004) develops this by arguing that individual and collective identities are social constructions that can only come into being within interaction. Furthermore, he suggests that because such interactions are in constant flux,

> (I)dentity can only be understood as a process, as 'being' or 'becoming'. One's identity – one's identities, indeed, for who we are is always singular and plural – is never a final or settled matter. Not even death freezes the picture: identity or reputation can be reassessed, and some identities – sainthood or martyrdom, for example – can only be achieved beyond the grave.
>
> (Jenkins, 2004: 5)

Hallam *et al.* argue that because the body is a key site in which the self is realised, it is precisely towards the dead body that attention should be focused in order to examine the transformation of identity upon death:

> Diachronically unfolding biographies collapse and condense in parting reconciliations, eulogies, reminiscence, graveside relationships and the gathering and subsequent dispersal of anecdotes, clothing and memorabilia. Holistically and synchronically, fragments of personal identity may be gathered, sifted and recast.
>
> (1999: 5)

Davies likewise sees the rituals around death as key to the (re)construction of identity, stating, 'Most societies set about ritual performances which help make sense of the ending of a life by reconstructing the identity of the dead within some wider framework of significance' (2002: 8).

However, it is not only the identity of the deceased that is (re)constructed through these rituals. Mourners, too, undergo such changes, both as bereaved individuals and as part of their community (Howarth, 2000). Walter (1996) argues that identity construction is a fundamental part of bereavement; by (re)constructing the identity of the deceased person, not least through funeral rites, mourners (re)construct their own identities as well. As noted by Davies, 'At death identity is altered not only through the loss of figures who have served as sources of identity but also by the new responsibilities which the living must take upon themselves.' (2002: 5) As it is through ritual acts that 'the identity of the dead and their network of links with the living change' (Davies, 2002: 42), it is within such rituals that social networks, broken by the loss of a group member, are re-formed. Hallam *et al.* also highlight the intersubjective nature of the self, arguing that if 'the self has agency and that agency inheres in the self, then biologically dead people who continue as active agents in the lives of the living have retained a form of self' (1999: 151). With regard to mortuary ritual, they assert that morticians

> (W)ork to produce a visual representation of the living person which corresponds to a remembered image of the embodied self of the deceased. [...] In preparing it for viewing, they seek to decontaminate it while recreating the deceased's personal identity, and to present the body in a life-like fashion.
>
> (Hallam *et al.*, 1999: 126)

Hallam *et al.* further argue that this is how a dead body is transformed from a 'disordered and dangerous object' into an 'extension of self' (1999: 132–138).

This chapter extends existing research by examining the viewing of unexceptional dead bodies (meaning bodies that are not already marginalised, such as those of stillborn babies [Hallam *et al.* (1999)]; see also Chapter 5) in the unexceptional settings of the American viewing room or the English chapel of rest. Its underpinning argument is that through examining in detail the viewing scenarios in each setting, a broader understanding of differences between western mortuary ritual behaviour – so often homogenised – can be gleaned. To do so, a conceptualisation of the spaces within which mourners navigate these death rituals is important. Therefore the chapter describes the spaces in which dead bodies were viewed (or not) at Durnford's in England and Blake's in the United States. It then considers how the differences in viewing habits led to differences in how the identity of the deceased person was (re)constructed as part of the wider community.

Blake's Funeral Home (United States)

In order to hold a viewing at Blake's Funeral Home, mourners (typically the next-of-kin) would rent a viewing room within the funeral home, paying for the use of the space as one might a function room (this has also been noted by Smale, 1985). The viewing rooms at Blake's were described by the staff as being '*comfortable*' and '*homey*'; at the same time, the rooms were designed to be plain enough to allow the casketed dead body to be the focal point. When the room was set up for a viewing, the chairs were arranged in rows facing the casket, which itself was centrally framed within the room in front of a set of drapes. Floor lamps were placed on either side of the casket, and overhead spotlights further foregrounded the casketed dead body. If requested, or if the deceased person (or next-of-kin) was Catholic, a kneeling bench was placed at the foot of the casket, and an 'everlasting candle' was placed at its head. The flowers provided by mourners were arranged on either side of the casketed body so as to create a visual as well as physical frame around the casket. The main (and often largest) floral arrangements purchased by the chief mourners immediately surrounded the casket; the rest of the floral displays, provided by members of the deceased person's extended community, fanned out and away from the body (Figure 7.1).

At Blake's, viewings typically lasted for 1–2 days, and the public were welcome for general viewing between 2–4 p.m. and 7–9 p.m.

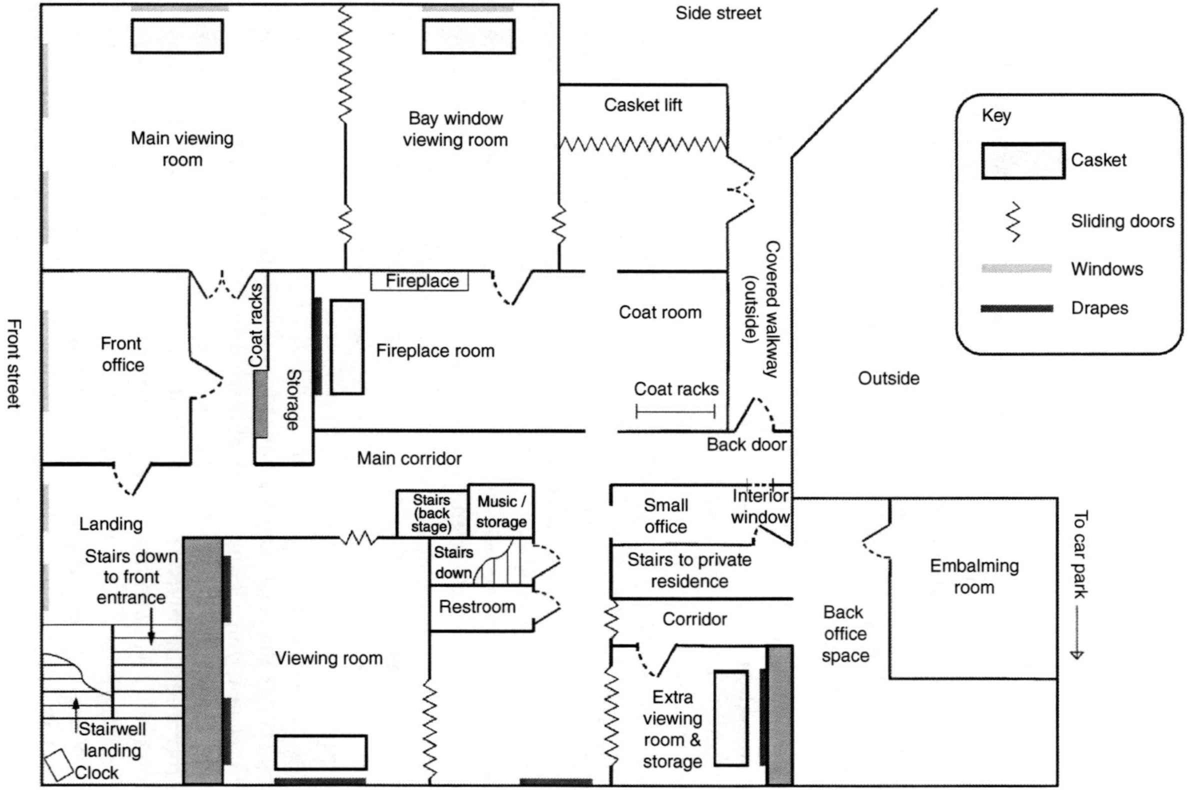

Figure 7.1 Blake's Funeral Home floorplan.

on one or both days. It was standard practice to announce the viewing times within the deceased person's obituary in the newspaper. For example:

> Mary (Smith) Jones,[1] 79, of Washington Place, died Tuesday, Jan. 3, 2006. Survived by her husband, Richard Jones; daughter, Susan (Travis) Schmidt, of Forest Heights; sons, Robert (Ellen) Jones, of Rupert Falls, and Ryan (Tina) Jones, of Washington Place; grandchildren, Richard, Shannon and William Jones, and great-grandchildren, Allison and Tom. Friends received from 7 to 9 p.m. Thursday and 2 to 4 and 7 to 9 p.m. Friday at BLAKE'S FUNERAL HOME INC., [address] where services will be at [time]. Saturday with the Rev. Martha Doran officiating.

Usually the family arrived 1 hour before the first general viewing so that they could first have some '*private time*' with the body. Non-family were not permitted to view before the family had arrived and had the initial viewing. However, once the initial viewing was complete (denoted by the principal mourner permitting the admission of extended and non-family), the viewing room became a public space. If a friend, neighbour or relative was unable to attend during stated viewing times, s/he might simply call in when it was convenient, for example, on their way to work. The presence of visitors outside set viewing times was common and often unbeknownst to funeral home staff because during the funeral home's opening hours the rooms could be accessed by anyone without guidance from the staff. This was not considered problematic or unusual by Blake's employees as the space was understood as public.

General viewing meant that for those outside of the immediate family, the body had a visible presence within the death ritual. Contrary to the theory that dead bodies have a destabilising effect on the social order because they signify 'the loss of self and the loss of individuality' (Howarth, 2001a: 120), visitors to the American funeral home actively incorporated the body into the ritual, *as a signifier of self*. This was manifest in several different ways. On entry to the general viewing, visitors might be guided up to the casket by the immediate family, or they might approach it themselves. Behaviour at the casket varied by individual and could include the following: resting their hands on the rim of the casket, touching the sleeve of the deceased person's clothing, touching the blanket, kissing the deceased person somewhere on the face, making the sign of the cross on their forehead, talking to the deceased person or simply looking at them without making physical contact. Mourners could also kneel on the prayer bench if one had been provided. The length of time spent at the side of the casket would also vary; if the casket was approached by a group (ranging in numbers from two to many), conversations might take place in the presence of the body.

Advertising the viewing in the newspaper meant that anyone who elected to attend the viewing could do so, leading to a range of people attending. In contrast with criticisms of the American funeral industry made by Bowman (1959) and Mitford (1963), mourners deemed the viewing experience valuable and important. This was not only important for the immediate family and very close friends; those within the broader community also understood attendance as significant because they saw the viewing room as a space where new social ties could be built and existing ones could be strengthened. Viewings were often the site of renewed relationships, where people who had not seen each other for many years could reunite and 'catch up' with one another. A viewing also provided a space for people from different realms of the deceased person's life to come together and have the opportunity to meet each other and share stories (see Walter, 1996), thus constructing a broader understanding of themselves within the social network of that person, who was also made present via interactions with their body.

Following the set period of general viewing, often attended by many dozens of mourners, the funeral would then take place. On the day of the funeral, mourners would first gather in the viewing room. Approximately 1 hour prior to the funeral service, *'final viewing'* occurred whereby the funeral director invited everyone to approach the casket so they could *'pay final respects'*. Pallbearers were typically invited first, followed by distant relatives and friends. Members of these groups formed a queue and filed past the casket, each spending various amounts of time there either individually or in small groups, often demonstrating the types of behaviours noted earlier. They were then guided out of the viewing room, leaving the immediate family in the viewing room with the deceased person's body. The doors were shut and the immediate family was again given some *'private time'*. Although families were permitted to govern how much time they could take, if the funeral service was to occur elsewhere (e.g., at a church), the funeral director would often encourage the family to *'say their last goodbyes'* and exit the room. Once the family had exited the room, the casket was sealed (closed) and then wheeled out to where the pallbearers were waiting to guide the casket to the hearse. The mourners would follow the casket and get into their vehicles, forming a funeral procession to the church. If the funeral service took place at the funeral home, the main viewing room was transformed into the '*chapel*' by the addition of rows of chairs and a lectern for the minister or officiant. In these cases, and if the viewing had taken place in another viewing room, mourners were guided into the chapel; the casket was sealed and wheeled out to the small sitting room while the service was taking place. Once the service was complete, the pallbearers guided the casket out to the hearse and the mourners followed.

If the deceased person's body was being cremated, the services terminated at the funeral home and the mourners either dispersed or attended the funeral luncheon, if one was being held. If the body was being buried, the mourners travelled in a funeral procession to the cemetery. At the cemetery there would be a short committal service in the chapel, normally lasting for approximately 5 minutes. It was less common, though not rare, for the family to '*go graveside*', for a service lasting between 5 and 10 minutes. Mourners did not observe the casket being lowered into the ground; when the service was complete, the casket remained above ground, resting on the metal rack that held it in place above the grave, until the mourners have departed.

(Re)Constructing identity at Blake's

Core to the American liminal period was the dead body, displayed in the funeral home for several days for all to see. Viewing the body at Blake's was thus a community-based, public experience. The 'success' of a viewing was commonly understood to be indicated by the number of those attending – the more viewers, the more successful the viewing, and the more the deceased person was (re)affirmed as a valued member of the community.

The inside of the (open) casket provided a highly decorated space that framed the body for viewing. Although its outside might be beautifully crafted, it was the casket's inside, with its plush upholstery, that was the focus of the community's attention as they interacted with the deceased person at the public viewing. Many of the casket's inner features – blanket, pillow, back panels – were purely decorative, and designed to enhance the body's public display.[2] The casket might further be filled with objects, placed there by mourners and personal to the deceased person, but often removed before the casket was sealed. This rendered such objects specifically decorative (i.e., part of the visual display at the viewing) as opposed to items accompanying the deceased person to wherever they were believed to be going. Furthermore, the highly decorated open casket that framed the body was itself located within the larger frame of the viewing room, which was also personalised with photographs, images or other objects. Such displays, representing different aspects of the pre-death self, contributed to and encouraged the active (re)construction of the deceased person's identity. In the American context this was designed to enable mourners to 'share memories' of the (pre)deceased person, thus leading to a shared (re)construction of the *deceased* person (Klass *et al.*, 1996).

Once the viewing ended and the casket was sealed, the dead body was no longer visible and became far less central to the ritual. As stated, the sealed casket was frequently not present at the funeral service, particularly if the service was taking place in the funeral home. The domed shape of the casket lid meant that nothing could be placed on top of it; thus the casket became an entirely practical vessel. Even while it was being 'carried' by the pallbearers, its figurative and literal weight were diminished by the fact that it is not *physically* carried but guided along on a wheeled trolley. Once placed in the hearse it could not be seen by onlookers as American hearses, unlike British ones, do not have glass sides. Furthermore, the floral displays denoting social relationships via the flower cards and placement at the viewing were absent thereafter. Only the immediate family's floral arrangements accompanied the casket outside of the viewing room; the rest were taken home by the mourners, returned to the florist or otherwise disposed of. The flowers that did accompany the casket out of the funeral home denoted social identity via their decorative ribbons, such as 'Wife' or 'Brother', rather than self-identity, such as 'Mary' or 'Richard'.

The funeral services observed at Blake's by and large contained prescriptive religious features rather than uniquely personal elements that contributed to the community's (re)construction of the deceased person's identity. For example, there was usually no specifically selected music aside from hymns, and if the service was taking place in the funeral home there was typically no music at all. There was, however, often a personalised eulogy or a section of the service during which attending members of the congregation could 'share stories' of the deceased person. In addition to the single narrative eulogy, any member of the congregation wishing to do so was welcome to stand up and share in verbally reconstructing the deceased's identity, thus rendering it a social and participatory event (see Garces-Foley and Holcomb, 2006: 225). However, this possible sharing of stories only aided in identity (re)construction if (a) the service allowed for it (which was not always the case) and if (b) members of the congregation were willing to participate (which they did not always do). Indeed, sharing stories and memories depended on the willingness of those present, rendering some services more 'successful' than others.

Finally, disposal at Blake's was perhaps the most invisible stage of the ritual process, because mourners did not actually observe it. If the body was being cremated, mourners departed the premises (the funeral home or church), leaving it behind for professionals to contend with (see Davies, 2002: 77). If the body was being buried, mourners tended not to witness this so the casket was left either sitting atop an open grave or lying within the cemetery's chapel. The final stage of the deceased person's transformation was thus neither witnessed by mourners nor formed part of their shared experience. While the liminal period at Blake's began with the body fostering the (re)construction of the deceased person's identity during the public viewing, as the ritual progressed, the body contributed less and less to the (re)construction of identity as it became increasingly removed from the ritual, until its final unobserved disposal took place.

Durnford Funeral Directors (England)

In contrast, the physical layout of Durnford Funeral Directors differed significantly from that of Blake's Funeral Home. There were seven chapels of rest in Durnford's main branch, which the staff described as being *'dignified'* and *'peaceful'*. Each chapel had a door that could be closed, and this door was kept closed whenever a body lay within that particular chapel of rest. Within Durnford's, the chapels of rest were located through a sitting room and along a corridor not readily accessible to the public (Figure 7.2).

The chapels of rest were not usually personalised. Infrequently, a framed photograph was supplied for the funeral directors to place in the chapel; this photograph would accompany the coffin upon its departure from the premises. Floral arrangements were normally delivered on the morning of the funeral and placed in the chapel of rest until the cortège was prepared. As they did not form part of the viewing display, they were not placed in the chapel in any particular formation, as at Blake's, but were simply stored there until they could be arranged in the hearse. All floral arrangements accompanied the coffin in the hearse to the church, crematorium chapel or grave.

It was Durnford's policy that all bodies were to be placed in a chapel of rest for the entirety of the day preceding, and up to the point

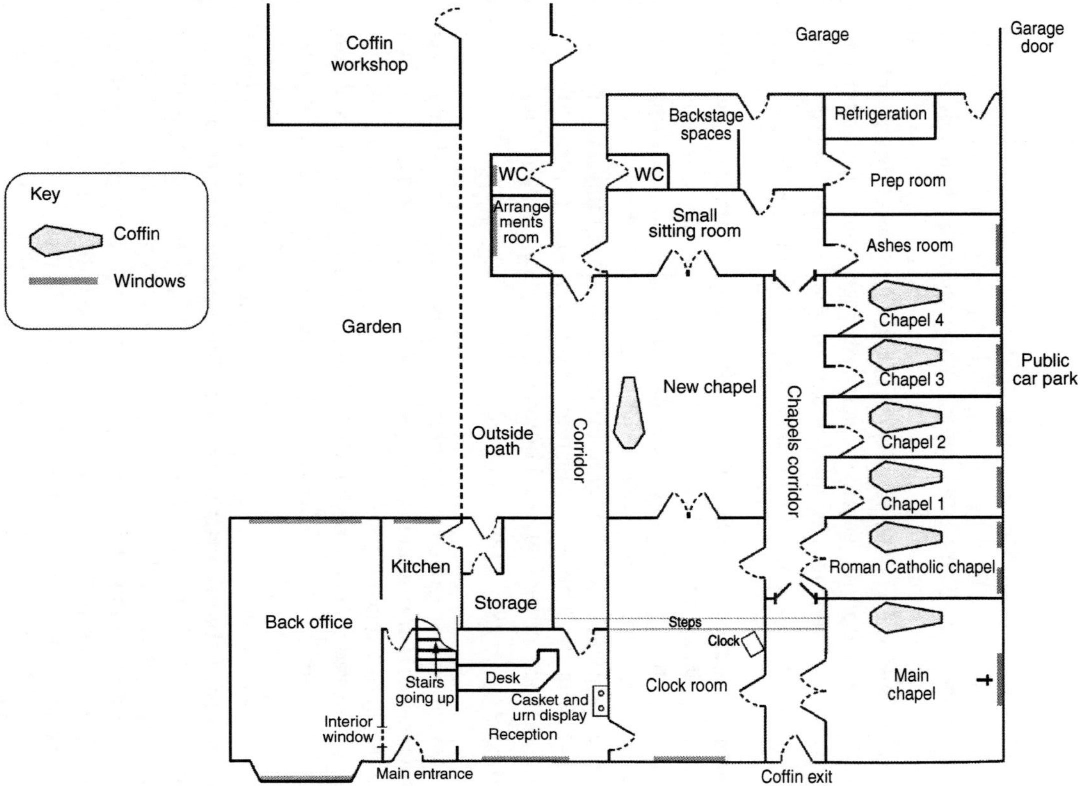

Figure 7.2 Durnford Funeral Directors' floorplan.

at which the cortège departed for, the funeral; this was regardless of whether the coffin was open or closed, or whether anyone came to view the body. Viewing was not a public event, so death notices made no mention of it:

> BROWN (nee SHAUGHNESSEY) (MAUREEN) – On September 16 2006 in hospital after a brief illness and of [area], Maureen aged 84 years, the beloved Wife of the late Layton, loving Mum of Laura, Roger and Derrick, a dear Mother-in-Law and a much loved Grandma. Service at [Church] on Friday September 22 at [time] followed by committal at [the] Crematorium. Family flowers only please but if desired donations may be sent to The Animal Sanctuary c/o Durnford's Funeral Home, [address], to whom all enquiries may be made [Tel].

Viewing at Durnford's was far less ritualised than that at Blake's. Mourners would typically just call in to Durnford's, possibly having telephoned first, and ask to view the person. Because viewings were not formalised (within set viewing times), the hours within which it was possible to view the body were the opening hours of the business: Monday to Friday 9 a.m.–5:30 p.m., and Saturday morning 9 a.m.–12:30 p.m. Anyone wishing to view outside of these hours had to make a separate appointment with the on-call funeral director so the premises could be opened and staffed.

When a mourner arrived at Durnford's to view, they were asked to wait in the reception area while the chapel of rest was checked. If the coffin was not yet in a chapel it was wheeled into one that was available. The lid was removed and propped in the corner. The body was checked and generally tidied (clothing straightened, head straightened, arms positioned) for presentation. The chapel light was switched on and a candle was lit. The door was left open, and the staff member ensured that all other occupied chapel doors were shut. The staff member then invited the mourner to follow them through to the chapel and, once guided to the correct chapel, left them there. The mourner usually remained there between 5 and 15 minutes, sometimes with the door open, sometimes closed. When the mourner returned to the reception, a staff member would speak to them prior to their departure; often the mourner, if they were responsible for the funeral arrangements, would have questions or points to raise. They might also comment on the body's appearance.

On the day of the funeral, how and from where the cortège proceeded varied. Typically, the coffin and floral arrangements were placed in the hearse; the hearse and lead car then travelled to a private residence – either that of the next-of-kin or that of the deceased person – where the mourners had assembled. If the cortège was departing from the funeral premises, mourners gathered there, at which point they may have had the opportunity to view the deceased person. The cortège then travelled either to the crematorium or to the church for the funeral service. Often, by request of the deceased person or next-of-kin, the cortège detoured from the direct route in order to travel past a particular site; for example, if the deceased person lived in a nursing home for many years, the cortège would pass the nursing home *en route* to the funeral.

If the funeral service occurred in the crematorium chapel, this was the final ritual that included the body. Once the service was complete mourners exited the chapel, usually chatting with each other and/or the officiant, and looked at the flowers and flower cards.[3] They would then either disperse or go on to the reception, if one was being held. Following a church service, the cortège travelled to either the crematorium or the cemetery where a shorter service would take place. At a burial, the coffin was ritually lowered into the grave in the presence of the mourners, although typically the grave was filled only after they had left.

(Re)Constructing identity at Durnford's

Identity (re)construction during the liminal period at Durnford's occurred differently to that at Blake's, as there was no communal public viewing. Viewing the body at Durnford's was a family-centred, private activity usually involving only the immediate family and possibly the deceased person's closest friends, who would typically come in ones and twos rather than in large groups. If disparate small groups arrived simultaneously to view the same individual, they would enter the chapel separately, one group waiting in the reception until the other left the chapel. Information regarding the opportunity to view was usually disseminated via the next-of-kin, who acted as gate-keepers to the body. Indeed, they could restrict viewing to only permit certain people; such restrictions, however, were rarely considered problematic by mourners or staff, as access by those outside the immediate family and closest friends was rarely sought.

The English coffin did not perform the same framing function as the American casket: its interior was far less decorated and it bore none of the luxurious fixings that make American caskets so aptly named.[4] Mourners frequently placed in the coffin objects personal either to the deceased person or to themselves; however, only those few mourners who viewed saw these objects *in situ*. Furthermore, it was rare for these objects to be removed prior to the coffin being closed, thus rendering them practical (i.e., to accompany the deceased person) rather than decorative (to form part of the visual display, as at Blake's). The open coffin therefore acted more as a receptacle for the body and its accompanying objects than a frame. As opposed to the American viewing room, the chapel of rest was rarely personalised beyond the coffin lid bearing the deceased person's name and birth/death dates. Once the coffin was closed, the deceased person's identifying details remained visible, thus personalising the coffin's exterior. Because it was flat, the lid also provided a surface upon which flowers or framed images of the deceased person's life could be placed. The latter, common in England, acted as an effigy that '[masks] the decaying natural body which [will] inevitably disappear, [and creates] a stable social body that [is] assured continuity' (Hallam and Hockey, 2001: 52). A framed photograph of the deceased person in life placed atop the closed coffin typically accompanied the hidden body in the hearse and at the funeral, serving to reinforce the continuity of the deceased person's social identity throughout the ensuing ritual. What is more, the coffin, with its personalising regalia atop, was visible to the general public as the glass-sided hearse travelled through the community on its way to the funeral service. This turned the coffin into a kind of transi tomb (Hallam *et al.*, 1999: 30–31): although the decaying and polluted natural body was not visible and was therefore left to the imagination, the social and idealised sacred body remained visually available.

In contrast to Blake's, although the liminal period at Durnford's began with private viewings that excluded the broader community, as the ritual progressed it became increasingly inclusive. The community only participated in the (re)construction of the deceased person's identity once the body itself was no longer visible, but had been replaced by its representative: the closed coffin displayed in the glass-sided hearse as it travelled through the community – often via the family home or other meaningful locations.[5] Floral displays provided by mourners accompanied the coffin to its final destination and were left for perusal post-service. As well as being accompanied by personal notes, the displays themselves were often personalised to reflect aspects of the deceased's social identity (e.g., '*Nana*') and self-identity (e.g., '*Mary*'). Although similar to services at Blake's insofar as they often included a personalised eulogy, those undertaken by Durnford's typically included specifically selected pieces of music (both secular and religious), reflecting aspects of the deceased person's identity.

Finally, methods of disposal at Durnford's were more visible to mourners than those at Blake's: although few mourners typically observed the coffin actually entering the retort, mourners did accompany the coffin to the crematorium. If the body was being buried, mourners watched as the coffin was lowered into the ground and sometimes actually participated in its lowering. Once in the ground, mourners would frequently interact with the coffin by dropping earth, flowers and/or other items onto it. In this position the plaque on the lid, bearing the deceased person's name, birth and death dates, remained visible to mourners. Mourners at Durnford's therefore shared in the deceased person's final transformation, invisible though the body might be.

Conclusion

This chapter has illustrated that in conjunction with – or perhaps because of – differences in American and English viewing practices, other aspects of the death ritual show marked variation. In particular, the inclusion and function of the English coffin or American casket, the flowers, the cortège or funeral procession, the funeral service and disposal all vary between the two settings and serve different social and symbolic purposes. A comparative approach shows that at Blake's the deceased person's body plays progressively less of a role in their identity (re)construction as the liminal period progresses, while at Durnford's the identity of the deceased person is progressively revealed and shared via the more visible role that their body – within its representative, the coffin – plays within the liminal period. Both rituals have public and private features that include or exclude the wider community: whereas at Blake's the broader community participate in identity (re)construction through rituals including the visible body in the funeral home, at Durnford's the opposite occurs, and there is a progressive participation by the community around the 'transi tomb' once the body itself is no longer visible.

By focusing on mourners' interactions around and including the dead body during the mortuary ritual, this chapter builds on the research of Hallam *et al.* (1999) and Hallam and Hockey (2001) in showing that the dead body cannot be understood as uniform. The differences in viewing habits mean that (re)constructing the deceased person's identity occurs differently in the two settings. At Durnford's in England, the deceased's identity is increasingly manifest as the death ritual progresses; at Blake's in the United States, the deceased person's identity is increasingly obscured as the death ritual progresses. Thus, the two rituals include the broader community in different ways, allowing mourners to partake in (re)constructing the deceased person's identity at different points of the respective ritual process. Valentine argues that, when considered within the larger context, modern death rituals 'make sense within the western cultural context of individualism, materialism and commercialism' (2007: 29; see also Kellehear, 1984). I suggest that within this 'western cultural context' there exist nuances that cannot be homogenised, as cultural practices vary markedly between, and indeed within, different western countries. These variations are rooted in everyday cultural practices, rather than in the differences of institutional ownership emphasised by Walter (2005). Although the two research sites are place-specific and may not therefore be generalisable nationally (Miles and Huberman, 1994), the evidence presented here suggests considerable scope for comparing and contrasting western societal responses to mortuary ritual through detailed examination of the spatial and temporal rituals that take place within the liminal period.

Notes

1. Names used within this chapter are pseudonyms.
2. Indeed, Habenstein argues that, even prior to the development of modern embalming, it was early American casket salesmen – *not* funeral directors – who propagated the 'value' of viewings, as the prepared and displayed corpse provided them with the opportunity to show off their wares (1954: 127).
3. Many crematoria have floral display areas outside, such as metal racks or allocated spaces on the ground, where the flowers from each service can be displayed. Often flowers from several different services are displayed alongside one another.
4. The word 'casket' traditionally means 'jewel box' (Habenstein, 1954: 127).
5. The London funeral directors Howarth worked with believed that 'returning the deceased to the place they knew as home was "only right and proper"' (1996: 177).

8
Private Grief in Public Spaces: Interpreting Memorialisation in the Contemporary Cemetery

Kate Woodthorpe

> Cemeteries are places where the living will go; cemeteries are places where our heritage is stored; cemeteries are places where nature, art and learning can thrive; cemeteries are meeting points for the bereaved and recreational visitors alike; cemeteries are part of our memory bank. Our cemeteries are cultural landscape; cemeteries are open air museums and spiritual places.
>
> (Hussein, 2006: 16)

Much more than a static landscape of headstones, benches and rose gardens, the contemporary cemetery is a dynamic space filled with assumptions, activities and perspectives, some of which are contradictory. This chapter sheds light on these contradictions through examining the material culture of the cemetery landscape, particularly at the site of the grave. It does so based on the proposition that the cemetery is a space in which private emotion (grief) and public behaviour (mourning) intersect in potentially problematic ways.

Using data generated from an ethnographic study of the City of London Cemetery and Crematorium (CLCC) in Newham, East London, the chapter's scope is twofold. First, it seeks to demonstrate the competing demands and expectations at play regarding the material culture of memorialisation in the cemetery. Challenging the assumption that the contemporary cemetery is a place of only 'sorrow and solace' (Bachelor, 2004), and using memorialisation as a focus, the first half of the chapter reveals the cemetery as a contested and dynamic space (see also Francis *et al.*, 2005). In providing further evidence of the source of this dynamism and contestation, the second half of the chapter considers activity around the graveside to explore how visitors and staff have differing expectations about what constitutes 'appropriate' memorialising behaviour that can stand for 'the social expression of grief' (Howarth, 2007: 225). Overall, the chapter's aim is to generate greater insight into analyses of grief as publicly expressed through the act of memorialisation through drawing on established grief models and their lay interpretation to demonstrate how different expectations about what grief constitutes manifest in material outcomes at the graveside. Within this, a normalising discourse is identified as core to memorialising activity within the site of the study.

Cemeteries today: An overview

This chapter (and others in this collection) contributes to a growing body of literature produced over the last decade on death, objects associated with it and the spaces in which it is encountered (Hallam and Hockey, 2001; Gibson, 2008; Sidaway and Maddrell, forthcoming). Within this, academic publications concerning cemeteries have grown in number coming from, among others, anthropologists, sociologists, archaeologists and landscape architects (see, e.g., Worpole, 1997, 2003; Rugg, 2000; Tarlow, 2000; Bachelor, 2004; Francis *et al.*, 2005). In the United Kingdom, research into cemeteries has also expanded, motivated by a range of factors including a shortage of space available for burial resulting in the introduction of grave reuse (Stirling, 2009) and concern over the management of older cemeteries (Dunk and Rugg, 1994), particularly those that have become a setting for deviant behaviour (Browning, 2004). Owing in part to the legacy of the Victorian commitment to burial in perpetuity (established through the Burial Act 1857, s. 25), these motivations are all real issues within the provision of cemeteries across the United Kingdom, where there is mounting concern about their sustainability (Hussein and Rugg, 2003; West, 2006). Over the years this effectively has meant that once space for new burial runs out, a site cannot generate any new income for maintenance and, without this, becomes vulnerable to neglect or abandonment by its owners.

In the last 10 years there have been moves to address the sustainability of cemeteries in response to these escalating problems of neglect and abandonment, pressure on space for burial and the increase in instances of deviant behaviour (see Environment, Transport and Regional Affairs Committee, 2001). Motivated in part by accidental deaths in cemeteries around the United Kingdom as a result of poor health and safety (see BBC, 2003), this impetus has generated several central government investigations into the state of cemeteries in England and

Wales (e.g., the Home Office, 2004). Many local authorities have subsequently embarked on 'make safe' campaigns that typically involve testing memorials and either fixing ones that are unsafe or laying them down. In cemeteries that are still active (i.e., accepting new burials) local authorities and cemetery managers have also become keenly aware of the health and safety implications of the type of mementoes left on graves by bereaved visitors (Morris, 2009), with many imposing stringent regulations on what is and is not allowed in the site. These safety-driven activities have been complemented by publications from organisations such as English Heritage (2007), where there has been a growing call for the recognition of the socio-historical importance of cemeteries and the need to conserve the Victorian aesthetic in some (see Rutherford, 2008, for a good overview of Victorian cemeteries in the United Kingdom).

The pressures from ensuring sites comply with health and safety laws, alongside the appeal for the conservation of some Victorian cemeteries, in many cemeteries across the country (such as at the City of London Cemetery and Crematorium, 2005), have led to the creation of conservation management plans, and a corresponding clamp-down on present-day memorialisation. Typically this has included policies on the type of headstones permitted and – the focus of this chapter – the more transient mementoes and artefacts that are left on graves: flowers, cards, food, toys and so on. Usually prohibiting visitors from leaving glass mementoes, metal objects and food, the enforcement of these policies in some cemeteries has meant that the grounds have been stripped of the artefacts that one would expect to see in a cemetery (e.g., wreaths at Christmas, plastic flowers). In some parts of the country bereaved visitors have expressed feelings of aggravation in response to such practices (see, e.g., Baker, 2009). In explaining why bereaved people might be so upset about artefacts being removed from the grave, Rosenblatt argues that the practice is like

> (K)illing a deceased person again, like being disloyal, like giving up memories of good times, like saying they do not care about the deceased, and like other actions that deny the value and importance of the deceased and personal connections to the deceased.
>
> (1996: 53)

Therefore, it is perhaps unsurprising that in recent years the way in which memorialisation is policed has become a significant and contested focus of attention for visitors and staff in cemeteries (Morris,

Figure 8.1 Memorialisation at the City of London Cemetery and Crematorium.

2009). It is within this somewhat edgy climate that this chapter is situated (Figure 8.1).

The research from which this chapter originates was a 4-year ethnographic case study of the CLCC undertaken by the author (Woodthorpe, 2007) and co-funded by the Economic and Social Research Council (ESRC), City of London Corporation and the Institute of Cemetery and Crematorium Management. Located in Newham, East London (approximately two miles from the location of the London Olympics site in Stratford), the CLCC opened in 1856 to provide a service for all those living in the Square Mile of the City of London (see Lambert, 2006, for a detailed account of the cemetery's origins; see also Brooks, 1989; Mellor and Parsons, 2008). At over 200 acres it is one of the largest cemeteries in the United Kingdom and at the time of the research a staff of around 90 was working at the site – a figure simply unheard of around the rest of the country. Because of its physical size and staff resources, the CLCC is known nationally and is often credited as one of the most important in the United Kingdom (see Curl, 1980).

The intention of this ethnography was to extend previously published and unpublished work on this cemetery (Francis, 1997; Francis *et al.*, 2005) to shed further light on the range of activity within it, from the perspective of visitors and staff, and those who live nearby. I interviewed over 100 people for the project (divided into three groups of visitors, staff and the local community) and spent approximately

60 days over a 6-month period undertaking participant observation in the site. The data that are utilised here are mainly from talking to visitors at the graveside and members of staff who had extensive dealings with visitors in the cemetery (e.g., the office staff and members of the grounds staff team). In making public the contributions of participants in this chapter, the data used are entirely anonymised beyond identifying the contributor as a visitor, staff member or member of the local community.

The 'stuff' of the cemetery landscape

As Hallam and Hockey (2001) have argued, cemeteries are full of objects loaded with meaning. Permanent and 'fixed' (such as mausoleums and headstones), temporary and movable (such as the mementoes left on graves), these constitute some of the 'things' of material culture associated with death (Gibson, 2008: 5). However, as this chapter will demonstrate, at the CLCC the 'things' of memorialisation (specifically, the transient objects that are left on graves), what they are doing in the cemetery and what they might represent or communicate to others do not carry the same meaning to all visitors and staff. Instead, as data revealed, meanings associated with memorialisation in the cemetery were highly ambiguous and contested; located in a space that is public the mementoes left in the CLCC were frequently observed and evaluated by others, their meanings being interpreted and ascribed at will (see Potts, 2007).

The argument proposed is that much of the contestation surrounding memorialisation in the CLCC came from divergent assumptions about grieving behaviour and its public expression (here understood as 'mourning'). In particular, the grave site was a focal point of tension. I suggest that this was because the grave and what was left on it was the physical site of the intersection of emotion that would typically be expected to remain private (grief) with its public, substantive expression. How these mementoes were interpreted, however, was multifarious – mirroring divergent *academic* explanations for what memorialisation is and why it occurs: For example, a memento

> (F)acilitates relationships between the living and the dead. Highly volatile, it encompasses the power and vulnerability of the dead who are both revered and protected. This power not only demands responses from the living but can also be appropriated by them.
>
> (Hallam and Hockey, 2001: 90)

An alternative explanation comes from Holloway who suggests that

> Memorials offer a form of immortality for those who have died as well as the possibility of a continuing link between those who have gone and those who remain.... The common thread...is that memorials provide a focus for social transition and a psychological and spiritual link between the living and the dead.
>
> (2007: 160–161)

Elsewhere it has been argued that memorialisation is a powerful tool in terms of identity (re)formation, through the merging of mementoes and objects with memories of the deceased person, and a life-once-lived (see Okely, 1983; Seremetakis, 1994, cited in Hallam and Hockey, 2001; Ash, 1996). Bradbury has suggested that 'contemporary memorialisation represents a flourishing custom in which people find a welcome conduit for the expression of the many complex emotions of grief' (2001: 218). Further interpretations, particularly of older historical memorials, have considered ways in which they can provide way to honour the dead (see King, 1998), while others have questioned what 'honouring' the dead might actually mean (Kellehear, 2000).

Thus, memorialisation can serve a number of purposes: to mark the location of the deceased person; to continue connections with the dead; to provide a tangible focus; to 'honour' the deceased person; or to act as a tool through which people can communicate with others, both dead and alive (see Potts, 2007). The particular focus of this chapter is on the way in which memorialisation in this cemetery appeared to be frequently 'read' as a sign of how 'well' someone was grieving. However, I will argue that the reading was shaped by the way in which viewers expected grief to be expressed and for how long.

In making a case for this, it is important to briefly highlight some of the work that has already been published on the public expression of grief (public taken to mean in shared, communal spaces and in the presence of other people). In an earlier publication on the juxtaposition of the emotional state of grief and its public display, Walter (1997, citing the work of Elias, 1985) argued that a loss of traditional ritual – such as mourning dress – had 'denied the old ritual forms in which strong emotion could be expressed in code, [as a result] mourners now find that emotional reserve forbids any public expression of grief' (p. 127). Indeed, Walter has been particularly vocal about the power of the expectation that bereaved people should grieve 'behind closed doors' (1990, 1991). Such is the influence of this expectation that the way that grief is expressed in public is typically seen as a *reflection* of how well someone is coping 'behind the scenes'. Mourning activity, as the social behaviour associated with grief and serving as an indicator of how well (or not) someone is grieving in private, can thus be subject to scrutiny.

Therefore, as it is here, memorialisation can be interpreted as a form of mourning activity that can be subjected to scrutiny and judgement. Bradbury (2001) notes that this is compounded by the way in which visiting a grave and leaving things can be a highly personal (and private) act but one that at the same time is extremely public. It is the tension between the assumptions about the private emotion of grief and the public nature of the cemetery landscape that, alongside health and safety requirements and the privileging of Victorian aesthetics, led to contradictory interpretations of behaviour at the graveside.

The context of memorialisation: Safety and aesthetics

As I noted earlier, there are significant health and safety pressures on memorialisation in cemeteries, which can profoundly shape the look of the landscape. The CLCC has not been immune to this, and as with other cemeteries across the United Kingdom, staff at the CLCC are legally tasked with ensuring that the cemetery is a safe environment for people to visit. Of necessity this objective can sometimes require the confiscation of potentially hazardous mementoes from graves. During my research, some of the objects that had been removed included glass vases and bottles, plastic statues and ornately crafted metal fencing. Food stuff, such as chocolates and biscuits, were also routinely removed because of the risk of attracting vermin. However, out of concern for how this might upset visitors who saw things being taken off graves, it was done first thing in the morning before the cemetery opened to avoid witnesses. Nonetheless, offending objects would often be replaced by visitors, thus creating a continuous cycle of mementoes being removed and replaced. Staff had their own theories about why this was the case:

> *That's their little memory to that person isn't it? They want to put on it what they want.*
>
> (staff member)

> *You've got to remember you've got thousands of people that own, as far as they're concerned, they own the grave. They don't temporarily lease it, the*

> *rights to it, they own that spot. So you've got lots of little plots within a big plot.*
>
> (staff member)

Data suggested that there were substantial difficulties related to the sense of entitlement that some bereaved visitors felt about what happens at the site of the grave (see Woodthorpe, forthcoming). This was confirmed by visitors who made comments about *'owning'* a private grave plot, when in fact they owned the Grant of Exclusive Rights of Burial to that private grave – effectively a lease over a limited period of time.

Uncertainty over entitlement was exacerbated by differing expectations, among both bereaved visitors and staff, of whether or not something that was left at the graveside was 'appropriate'. For staff charged with managing memorialising activity at the grave, the decision to remove mementoes made of glass, metal or food was quite straightforward as, in terms of regulations, they clearly presented a danger to visitors or were detrimental to the sanitation of the site. Beyond this though, there was often uncertainty about the extent to which something was regarded as 'suitable' for a cemetery setting that did not clearly contravene health and safety regulations. Examples of such more ambiguous mementoes included plastic figurines more commonly seen in garden centres (gnomes, angels, toadstools), small stone statues, wind chimes and solar-powered frogs that would croak at dusk. In these cases, more often than not the indication from staff was that they felt that their job was to regulate memorialisation based on health and safety grounds, so unless an object was a clear hazard, taller than the headstone or protruding beyond the headstone area (and therefore presenting a problem for mowing around the grave), they would usually leave the objects *in situ*.

However, the decision to leave some of these mementoes in place could also be a source of frustration for other visitors who thought that some of these objects were not 'fitting' in a cemetery landscape. Examples of mementoes that caused friction because they were left by staff but deemed by others to be offensive included stuffed toys (often vulnerable to decay when exposed to the elements) and balloons. Many of the criticisms made to me about these types of objects were articulated in terms of the disorder and disarray they created, conflating safety in the site with aesthetic value.

> *How can I say this? I think a lot of what is put on graves is really untidy, [it] makes them look a complete mess.*
>
> (visitor)

> *I like things to look neat and orderly. If I had my way there would be no rubbish on graves.*
>
> (visitor)

Thus, through leaving mementoes on graves – or not removing them – visitors and staff revealed contradictory expectations about what is 'fitting' in a cemetery (see Francis *et al.*, 2005). These expectations reflected different interpretations of the function of the cemetery landscape more generally. On the one hand, some visitors and staff who participated in this ethnography indicated that they expected the cemetery to be a beautiful and tranquil site that could be utilised for anybody's quiet contemplation and spiritual fulfilment, much like the original intention of many Victorian cemeteries in the nineteenth century (Tarlow, 2000), and espoused in publications such as *Paradise Preserved* (English Heritage, 2007). On the other hand, there were visitors and staff who felt that the cemetery should provide a space that privileged the needs and demands of bereaved people (particularly newly bereaved people), regardless of whether these conflicted with other people's desire for an attractive and peaceful landscape. This type of contradiction was particularly apparent in the use of wind chimes, for example, which caused substantial friction between those who liked the sound they created and others who found it intolerable.

This brief overview of the competing pressures on and expectations about memorialisation demonstrates how material culture of the cemetery environment is politicised via the (sometimes conflicting) demands of safety, aesthetics and the needs and demands of visitors (see also Bradbury, 2001; Francis *et al.*, 2001). Such varying requirements of cemetery space reflected assumptions that underpinned what constituted 'appropriate' memorialisation in the cemetery, a situation further complicated by different interpretations of how people publicly expressed their (private) grief.

Interpreting memorialisation in the cemetery

I have argued that pressures on memorialisation in the cemetery included concerns about its safety, aesthetic value and the extent to which these met, or conflicted with, the needs and demands of bereaved people. However, as my data revealed, bereaved people themselves did not share a common view of their own needs, which partly depended on the way in which they understood the concept of grief. The second half of this chapter considers this issue by examining memorialisation as the public, material expression of loss and offers two possible interpretations of memorialisation based on how people viewed bereavement, grief and relationships with deceased people.

Psychological models of grief and their uses

Possibly the most popular and well-known way of conceptualising grief – both in academic and in lay terms – has been the idea that a bereaved person goes through various emotional stages over a period of time, their transitions being shaped by a prevailing theme of 'working through' emotion (Grimes, 2000; Littlewood, 2001; see also Small, 2001 for a good overview of this perspective). These emotional stages include shock, anger, sadness and forgiveness and bear many resemblances to Kübler-Ross' work on stages of dying (1970). An interpretation that owes much to the work of clinicians and psychologists, this understanding of grief is as a process-orientated condition (Sanders, 1999) that blends 'emotional and cognitive reactions to loss' (Rosenblatt, 1996: 45). Any emotional response deemed 'excessive' is therefore potential evidence that someone is not managing their grief very well and possibly might need some kind of intervention. One consequence of this perspective is that grief is normalised in that it provides a way of distinguishing a 'normal' response to bereavement from one that is 'pathological'.

Although this approach to bereavement has been questioned by some sociologists because of its perceived neatness and exclusion of social context (see Howarth, 2007), explanations of the feelings and responses associated with bereavement has been popularised via what has been termed 'clinical lore' (Walter, 1999 drawing on Worden, 1991). At its most basic, clinical lore is the way in which lay people utilise psychological ideas about grief to make sense of their own, and others', responses to loss. My data confirmed the popularity of this interpretation. For example, one staff member commented on a visitor who had been attending the site daily for almost 10 years and 'still *left stuff on the grave*' (original emphasis). For this staff member the visitor was going beyond what they considered a 'normal' frequency of visiting. Another staff member commented on how the same visitor '*should have stopped visiting [and leaving mementoes] so often by now because the death happened so long ago*'. A further staff member said they felt sorry for the visitor as it meant that '*they could not move on with their lives . . . as they keep coming back here* '. Accepting memorialisation as the material and public expression of grief, comments such as these support the idea that grief should not go on for *too* long (Walter, 1997).

Likewise, data already presented, which described the uncertainty among staff over the removal of mementoes from graves, indicated that the way in which memorialisation was 'read' was a process of ongoing negotiation rather than a simple application of rules or theories. Visitors and staff frequently speculated about how memorialising activity might indicate how 'well' someone was coping behind closed doors. For example, if visiting and memorialising were deemed either excessive or out of synch with the expected progression of emotion, they might consider this to be a sign that they were experiencing some kind of problematic grief. There were instances too where memorialising activity was taken as confirmation that the memorialiser was having a pathological or abnormal response to their bereavement (Hallam *et al.*, 1999). As a staff member speculated:

> *I do think they should kind of rein it in a bit, if you know what I mean? I mean, look at it, that person died almost twenty years ago, and they're still coming and leaving stuff on it?...I dunno, it just seems a bit too much for me.*

Another participant, this time a visitor, thought that excessively long memorialising activity was indicative of the memorialiser experiencing the emotion of guilt as part of their grief:

> *Yeah, sometimes it can go on for too long....I think it's because they're guilty or something.*

For those people who implicitly drew on clinical lore to make sense of memorialising activity, this interpretation carried powerful assumptions that there was a 'normal' way to express grief in public in terms of scope and time. It also suggested that the *way* in which someone memorialised in the cemetery was read as an indication of their state of mind (see Potts, 2007). There was a clear expectation that grief should in some way 'finish' and that memorialising activity at the graveside should correspond in a timely fashion, based on the assumption that grief was a straightforward passage through chronological time (Small, 2001). The memorialiser whose activities did not correspond with this assumption ran the risk of being perceived by others to be grieving 'strangely' or at odds with what was 'normal'. This interpretation necessarily required that the memorialiser show restraint in their activity, to avoid being seen as pathological or emotionally out of control.

Thus, in this project explanations for memorialisation from a psychological perspective were common. However, the danger inherent in a psychological perspective is its restricted focus on the individual with less attention paid to the wider social and material environments in which grief is expressed. The assumption that visitors acted individually without being influenced or shaped by their external environment meant that the focus was entirely on the psychological management of grief at the site of the deceased person's interment. In the rest of this chapter I offer an alternative theoretical interpretation of memorialisation which shifts this emphasis from emotional control to the relationship between the living and the dead.

Continuing bonds and memorialisation

Certainly, data from some visitors and staff indicated that the relationship between the living and the dead, partly played out in the physical location of the deceased person's remains, was very important to bereaved people when they were 'visiting' the dead (Woodthorpe, forthcoming). These visits often marked occasions such as birthdays, anniversaries and other annual events such as Christmas, and frequently included the bringing of a gift of some sort (which was left on the grave) (Figure 8.2).

Visiting and gift-bearing practices of this kind suggested that memorialisation was not simply a public expression of grief as psychologically based interpretations outlined earlier seemed to suggest. This alternative interpretation put forward by some visitors and staff resonates strongly with the idea of continuing bonds, a theoretical approach to bereavement that has moved away from the idea that grieving involves the severance of ties in order to move on in life, towards an incorporation of the social context and materiality of grief (see Valentine, 2008). As an interpretation of memorialisation at the graveside that focused on the relational context of bereavement, data from staff and visitors in these cases indicated a belief that bereaved people can, and do, have ongoing relationships with the dead that do not necessarily correspond with discrete, time-limited emotional responses, as a psychological interpretation would suggest.

An alternative 'reading' of memorialisation in the cemetery then was as the material manifestation of a continuing bond with the deceased person. Thus memorialisation not only ensured that there was a perceptible presence above ground for the deceased person (see Meyer and Woodthorpe, 2008), but it also served as an opportunity to bestow

Figure 8.2 Birthday celebrations at a grave in the CLCC.

presents upon that person and sustain a sense of them being 'cared' for. The idea of a relationship between the living and the dead played out through memorialisation was often articulated by visitors, as the following comments indicated:

> *I mean, I know they're dead already but to see it look in a big mess like that, it needs to be neater and tidy, and taken care of. Like someone's almost caring for them.*
>
> *I think for some people it's not just to do this, lay the flowers and that sort of thing. They find comfort, they do, they talk to them, say things, they ask their advice.*

> *We have a look at the stones, so we do see what people leave behind. There were lots of beer cans left behind on a lad's grave, I think it was his 21*st *and his friends must have come and had a drink with him to celebrate.*

An alternative analysis shows how memorialisation was viewed less as the public expression of grief and more as evidence of ongoing emotional attachment to the deceased person. Consequently, there would be fewer expectations about 'normal' emotional expression and its cessation after a certain period of time. Nonetheless, interpretation that allows for such continuity was not without its challenges. Much like the quote above where a visitor speculated on the possibility that someone was memorialising because they felt guilty, through observations and by questioning visitors and staff who speculated on individual's motives for memorialising, I argue that someone's behaviour was often interpreted as a representation of what the dead person meant to them:

> *I think it's possible to look at some of the graves out there and see what someone is thinking, I really do. And some of them show it better than others too.*
>
> (staff member)

On a number of occasions, staff and visitors hinted that they felt that these ongoing connections, materialised through memorialisation activity at the graveside, were as much about presenting some kind of lasting impression, or performance, for other cemetery visitors and showing other people that they still cared, as they were about the actual relationship between the living and the deceased person. For example, as the following visitor commented:

> *I come down here about once a month or so and leave something on his grave, to let them [my family] know I've been here.... And I think that's the right thing to do, just to show I still love me Dad.*

However, for visitors and staff, despite a belief that ongoing memorialisation activity was acceptable, there were still potent expectations about the need to show restraint in the public setting of the cemetery, and questions raised about how well someone was managing the (now changed) relationship (or suggestions that they had had a messy relationship with the deceased person) when that memorialisation was deemed 'excessive'. Thus, although freer from constraints about how long memorialising activity should last, for those who recognised memorialisation as a demonstration of an ongoing bond with the deceased person, the material manifestation of that continuing bond was subject to scrutiny and judgement.

Conclusion

In this chapter I have questioned the extent to which it is useful to think about memorialisation in terms of whether it is 'normal', 'abnormal', 'good' or 'bad'. Although this type of crude distinction might paint an overly simplistic picture of memorialisation, the evidence presented in this chapter suggests that there are powerful normalising discourses in the cemetery about what constitutes 'fitting' memorialising activity. Some of these discourses are upheld and endorsed by local authorities (health and safety) and national interest bodies such as English Heritage (Victorian aesthetics). However, there are also potent normalising discourses associated with expectations about mourning practices in public spaces and their material manifestation. This resonates with Walter's assertion where,

> The norm is not just that you should grieve in private, but that others should know you are grieving – otherwise you might not have cared for your husband/mother/child after all!
>
> (1997: 133)

Data for this project revealed that visitors to the CLCC could find themselves in an ambiguous position whereby to memorialise *too* much was read as a signal of poor grieving; yet to not do it *enough* was seen as an indication of not caring 'enough'. For some visitors and staff, there was an expectation that memorialising activity would tail off as the bereaved person moved through different emotional 'phases' of grief; for others it would be appropriate to continue indefinitely memorialising, to recognise significant dates and continue a bond with the deceased person. In the public space of the cemetery, it was the culmination of these varying perspectives that resulted in friction about what was left on graves, and why.

This friction confirmed the ambiguous purpose of memorialisation noted earlier in this chapter: to mark the site of the deceased person, or to continue a bond with a deceased person and reaffirm their presence, while at the same time representing a means of communicating with others (Potts, 2007). The 'public-yet-private world of the cemetery'

(Francis *et al.*, 2005: 107) necessarily meant that memorialising activities were subject to institutional practices (such as the removal of mementoes on health and safety grounds) and open to scrutiny by others. Further, the requirement to fulfil so many potential functions, derived from discourses of health and safety and aesthetic value, made conflict over the way in which memorialising activity in the cemetery was interpreted almost inevitable, as seen in tension surrounding the use of gnomes, wind chimes and solar-powered frogs.

If memorialisation in the cemetery was therefore perhaps inevitably contentious, those who managed the site – the cemetery staff – were consequently placed in a challenging position. Negotiating competing and, at times, conflicting expectations of and pressures within the site, cemetery staff were tasked with avoiding alienating current visitors, while facilitating an environment that caused the least amount of hostility. This was no mean feat, and there is much more that can be learned about the way in which they do this. There is also much more to be considered in terms of contextualising memorialising activity with regard to age, gender, ethnicity, religious belief and so on. Beyond the scope of the research from which this chapter originates, this context is part of the memorialisation picture and would benefit from further inquiry. The empirical evidence presented in this chapter certainly suggests that the material culture of the cemetery is a fruitful area for study and there is much potential for future exploration of memorialisation, its function and its associations with theories which promote particular forms of mourning and grieving. As the chapter has illustrated, there is still much to be gained from examining the relationship between death, emotion and place.

9

Wandering Lines and Cul-de-sacs: Trajectories of Ashes in the United Kingdom

Leonie Kellaher, Jenny Hockey and David Prendergast

Cremation was introduced in the United Kingdom in the 1870s and legalised in the 1902 Cremation Act (Jalland, 1999). Its uptake was slow however and only overtook burial in 1968. By 1995, just over 70 per cent of all disposals in the United Kingdom were cremations, the rate then remaining at this level (Davies and Mates, 2005). As a result of this relatively low initial uptake, local councils were unwilling to invest in new sites for crematoria, instead using existing cemetery land. In addition, their location in existing cemeteries was seen to 'mitigate the "strangeness" of this new form of disposal by offering its provision in a familiar context' (Rugg, 2006: 217). While the siting of crematoria subsequently became more diverse, cremation and the disposal of ashes usually occurred in the same landscape: the crematorium and its surrounding gardens. Thus in the 1970s only 12 per cent of ashes were *removed* from crematoria by family or friends for disposal elsewhere (Davies and Guest, 1999).

However, by 2005 practices had changed and on average 60 per cent of ashes were being removed. Our ESRC-funded project,[1] *Environments of Memory*, asked where those ashes were going and what practices might surround their disposal. In Nottingham, London (Barking and Dagenham), Glasgow and Sunderland, we interviewed minimum samples of 15 bereaved people who had removed ashes and seven death care professionals per city (see Prendergast *et al.*, 2006). In addition to this some crematoria provided supplementary quantitative records of ashes strewn in their Garden of Remembrance. Unlike the 40 per cent of people who leave ashes to be disposed of in this way, with or without their attendance, those who remove them must decide on, and find, somewhere to put them. We wanted to know how their choices, the locations selected, the ashes themselves, and anyone else involved, came together in particular acts of disposal and what might be the consequences of these acts.

Therefore, this chapter explores the decision-making processes which led family members or friends of someone who had died to choose a particular mode or site for the disposal of their ashes. The deliberations involved place and space, yet the temporal dimension of these processes was also important; many interviewees needed time to arrive at choices that 'fitted', or made sense, in terms of the person they had known and their previous and future relationship with them. Finding the proper 'fit' could, nonetheless, result in final destinations that were akin to traditional practices; for example, interment in existing family graves or ash graves in cemeteries (see Kellaher *et al.*, 2005). The private home or garden could also become a destination, although the interior of the house tended to be chosen for temporary storage until a decision had been made or another set of ashes – usually a spouse – could be mingled with the first for final deposition. 'Nature', as in water or landscapes that held special memories, was another choice, along with less contemplative settings such as football grounds and pubs. While the media often flag the exotica of ash disposal – being smoked by a rock star, fired into space, made into a diamond, incorporated into a painting or a piece of sculpture – we frequently found mundane destinations for ashes that were anthropologically more revealing since they carried associations with lives previously lived, whether everyday life or key life course transitions such as weddings. Many interviewees seemed to have worked and re-worked such associations, using material aspects of a former shared life – places, events, odd moments and particular objects – to craft particular places in which the deceased person's habitual dispositions could persist in death.

Using material culture studies' perspectives, this chapter explores these data as examples of place-making. To this end, Ingold's (2000) work on the relationship between people and the landscape is helpful, as is his subsequent work on line-making (2007). This later study examines the lines people generate wherever they go, whether through walking, talking or gesticulating. Key to Ingold's thinking (2007) is the difference between 'open lines' which involve unpremeditated and creative movement towards unidentified destinations – of which wayfaring is an example – and 'closed lines' which lead only to a fixed endpoint, exemplified in commuter travel. This distinction provides a theoretical resource which allows comparison between the journeys made by people who retain ashes without a destination in mind and an established route between the crematorium and its adjacent Gardens of Remembrance. While Ingold's discussion includes a breadth of examples that includes knitting, handwriting and musical notation, we focus here on lines that somehow 'mark' the landscape, even if invisibly, along with the spatial destinations that represent their end points.

Environments of memory

Many interviewees said that disposing of ashes in a way they felt happy with had been challenging, referring to anxieties about misjudgements and an 'unfitting' – and generally irretrievable – disposal. This emphasis emerged from the data themselves, rather than in response to particular interview questions, many people describing how long it had taken to decide what to do and the constraints they had encountered. As members of UK society, interviewees in fact had an extensive range of choices yet the relinquishing of ritual and moral claims on death and its aftermaths by ecclesiastic and other authorities (Rugg, 1999) meant they faced a vacuum in terms of what was acceptable and 'proper' practice. Unlike many European countries, the United Kingdom barely regulates ash disposal (Arber, 2000); interviewees nonetheless often imagined that what they had done breached unspecified legal or social boundaries leading them to act covertly, sometimes under cover of darkness. Occasionally, this subterfuge was subsequently incorporated into the meaning of the event. For example, when a humanist officiant had been refused District Council permission to scatter a gay friend's ashes in a public area where he would meet other men, the officiant just did it discretely, despite feeling *'very illicit and very dodgy'*. Afterwards, though, he said: *'But then I thought well what he did round these trees was fairly illicit and dodgy ... he probably would approve'* (see Hockey *et al.*, 2007a). Other interviewees described the surprise or uneasiness of visitors when they had placed ashes on display at home but in some cases left them there in quiet defiance of other people's response.

These data led us back to van Gennep's ([1909] 1960) rites of passage schema, in some ways an obvious resource for interpreting the progression of ashes from a crematorium to their final resting place. Thus cremation might be seen simply as the first stage in a process that moves through a liminal period of waiting or indecision, culminating in a third stage, the disposal of ashes at a site eventually decided upon. That said, a sizeable minority of people in our dataset still chose cremation for its simplicity and finality, and were satisfied with disposal of ashes in Gardens of Remembrance immediately after the cremation, possibly never re-visiting the site. For example, data for a single calendar month on the destinations for all cremated remains at a London crematorium showed that the ashes of older people were more likely to be strewn at the crematorium, without any witnesses present, making this a one-stop process. This response was summed up by an interviewee who said '*The curtains close, coffin sinks down – that's final*'. For younger groups, however, it appeared that there were more surviving family members to take the ashes away for private disposal later. They were therefore more likely to feature among the interviewees for whom a second 'burial' in the form of a final ash interment or scattering not only ended a liminal phase that might have gone on for years but also completed the material disposal of the deceased person.

In this respect van Gennep's rites of passage schema is telling in terms of the shifting social status and location of the dead. Interviewees described struggling to make a place for them in a society that no longer offers a taken for granted religious framework for envisaging where the dead will be, in whose company, and with what purpose. When people described the period between cremation and final disposal it became apparent that not only they, but also the deceased person, were passing through a liminal period within which that person somehow had to *settle into* the state of being dead. This notion of 'settlement' features within Hetherington's (2004) discussion of disposal in a broader sense, one that draws on Hertz's ([1907] 1960) work to describe coming to 'a settlement with how we manage our relations with others in terms of our memories, a sense of tradition, and through our relations not only with our contemporaries but also with our ancestors and future generations' (2004: 172). In our data, however, a final rite of reintegration (van Gennep, [1909] 1960) proved elusive as interviewees strove to understand where the deceased person might ultimately be headed and, indeed, their ontological status. They could thus be compared with members of the supposedly less individualistic worlds that van Gennep described when he envisaged *societal* rifts being repaired by final rites of passage. For our interviewees it was *familial* ruptures brought about by death that might be experienced as repaired (see Hockey *et al.*, 2007a).

Within an environment where both control *and* guidance about ash disposal by religious and legal authorities is minimal, our data raised theoretical questions about experiences of place-making and whether the final destination or place was constructed for the dead person, for bereaved people or for the members of some other category such as descendants – or indeed for all of these. Van Gennep's ([1909] 1960) rites of passage schema does not, however, address such questions and indeed his work has been criticised for not addressing the experience of transition (Hockey and James, 2003), a dimension missing from the theoretical agendas of his time. Subsequent work has asked how particular qualities of experience come into being (see Turner, 1969) – but this too has been criticised by anthropologists such as Csordas (1994, 2002) and Ingold (2000) for its failure to recognise embodied experience, or being-in-the-world, as the site of knowing.

Thus Ingold challenges the cognitive orientation of a social science belief that 'human beings inhabit discursive worlds of culturally constructed significance' (2000: 172). Instead he speaks of people's life-worlds, arguing that we understand our environments through our *being*-in-the world, by acting *on* and *in* the world. In other words, he questions arguments that suggest a culturally specific classificatory framework being imposed on a continuous and undifferentiated material environment – the idea that the shape and order of the world about us derive from our own cognitions. This critique is borne out in our data which evidence the shaping of *after-life*-worlds fit for co-habitation by the living and the dead in a common territory made up of familiar places and habits. Such arguments represent our imaginings, or mental representations, as indeed stemming from our being-in-the-world, from what Merleau Ponty (cited in Ingold, 2000: 186) calls the homeland of our thoughts. When we stand and stare, then, our feet are nonetheless firmly on the ground.

These perspectives can illuminate the thought processes and decisions that removing ashes for later disposal involves. In these circumstances, the matter, or materiality, of ashes, where they are located, both temporarily and permanently, does *matter* to people (see Prendergast *et al.*, 2006). People cared intensely if ashes were lost or withheld, if someone else chose their destination, if the disposal 'went wrong', if people felt coerced into disposing of ashes by someone they deemed closer to the deceased person (see Hockey *et al.*, 2007b), if *nothing* happened to the ashes, if they were left pending at the crematorium, and of course if they were scattered in the Garden of Remembrance against their wishes.

Thus while ashes have material properties that allow them to be scattered to the wind or dispersed within water, their *place* does remain important to people (Prendergast *et al.*, 2006), a concern which has an historical context. Certainly the United Kingdom's post-Christianisation tradition of churchyard burial meant that the buried body has not always had a permanent location or marker, churchyards often containing many more bodies than the names on gravestones indicate (Horrox, 1999). However, from the early nineteenth century, private and municipal cemeteries created a different and profitable association between burial and place. Guarantees of marked burial in perpetuity became a powerful selling point at a time of body-snatching on behalf of ambitious medics, when churchyards overflowed with corpses (Rugg, 2000). The importance of a marked location also echoed earlier fears of personal erasure in a common or pauper grave, and the religious exclusion of unbaptised infants, suicides and criminals buried in unconsecrated ground (Gittings, 1999). As Chapter 5 explores in relation to Flanders, Belgium, only since the 1970s has the grief of parents whose stillborn children were disposed of anonymously in hospital grounds or municipal cemeteries been acknowledged (Lovell, 1997).

This meant that, for example, for one interviewee, releasing her mother's ashes into a river meant that all water now allowed contact with her mother. A variant on this commitment to a materially located encounter is evident in data gathered by Springett in the early 1990s from the families of former naval personnel who ceremonially 'bury' their ashes at sea in a permeable casket (see Kellaher *et al.*, 2005 for a more detailed account of this study). For them, the nautical map that marked the grid reference of the 'burial' was an important point of contact with the deceased person. Grounded in individuals' shared lifeworlds, then, the placing of ashes by those who survive the deceased person is a process that involves the re-working of embodied memory within environments such as these; it is an act of place-making.

Place-making

To understand what constituted a fitting 'place' for the dead – and how a sense of 'fit' was arrived at and sustained – to account for the experience as well as the structure of this rite of passage, we draw here on Ingold's (2000) 'dwelling' perspective. Ingold (2000) asks how people make themselves at home in the world and queries a distinction between animals and human beings which assumes that the latter impose pre-existing mental representations onto an external environment, while the former enact hard-wired imperatives. If, as suggested, the world is the homeland of our thoughts, then human beings' mental representations arise within 'the current of their involved activity, in the specific relational contexts of their practical engagement with their surroundings' (Ingold, 2000: 186). If this is so and buildings, for example, are crystallisations of human activity within an environment, can this 'dwelling' perspective help make sense of the places within which the living position the dead? Ingold describes all place-making, whether by human or non-human animals, as occurring within environments modified by predecessors, by the material residues of previous spatial and temporal activity. So we are not only part and parcel of our material environments in the present; we are also bound temporally to those who preceded us, to *their* beings-in-the-world and to their after-life worlds.

This perspective helps explain how ashes, as the residue of particular individuals, lead bereaved people into disposal trajectories that somehow 'fit' the dead and the living (see Hockey *et al.*, 2007c). Ashes are made to operate both within and upon the places and things shared by deceased and bereaved people, so contributing to survivors' contemporary and anticipated memorial activity. Such choices can be located in their 'enmeshment within an all-encompassing field of relations' (Ingold, 2000: 187), a field that both precedes and flows on after them. Thus, one female interviewee described needing time to assemble the fine detail of a disposal that would mesh with the places and modes of the couple's past lives and leave her with something more than a recollection of an event.

> *I knew I wanted them back and I knew I wanted to put them somewhere and I knew I wanted it to be at the sea but I didn't (know) when or where or how. I just couldn't. I don't know if it was something psychological telling me, you know, not to do it yet. And I didn't know exactly which way I wanted them. I think if I'd done it at the beginning I wouldn't have really... Oh, I don't know how to explain. I just don't think it would have had any meaning – it would have been just a scattering of ashes. But I wanted to start to get on with my life and then I knew where I was putting him and why I was doing it. I think if I'd done something on the spur of the moment and hadn't had a really good think about it, it wouldn't actually have been what I wanted to do.*

Though confident about the main disposal event, she took several months to organise components such as the place and time, the people she wanted to be present and to assist. These details, she felt, would imbue secondary disposal in her local harbour with meaning that fitted with her husband's past life, so that there could be congruence with the ongoing context of relationships she would find sustaining.

Ritualisation and meaning-making

If, following Ingold (2000), we treat meaning-making as an engaged, open-ended process and not an event or occurrence fixed in time and space, then rather than a single death ritual, we are dealing with an open-ended process of ritualisation (see Seremetakis, 1991). Thus, for example, interviewees often navigated their route towards disposal in fits and starts, they had little sense of a comprehensive map, or they simply talked about what felt right or wrong. So meaning for them, like place, might not be bounded or pinned down. This contradicts twentieth-century medicalised models of grief which popularised the notion of 'closure', the need to render the dead into some form of manageable memory (Small, 2001). As Chapter 8 also details, such models have now been challenged, with the experience of continuing bonds gaining ascendancy by the end of that century (Klass *et al.*, 1996). This generates a view of meaning-making as participative, emergent and temporally indefinite, past, present and future shedding their linear relationship with one another; instead, through processes of ritualisation, they loop back to inform one another or indeed shift position as aspects of meaning-making processes.

The connections Ingold (2007) makes between place-making and line-making are helpful here. As outlined above, he differentiates between lines that 'go for a walk' in an active and open-ended sense, and those that connect fixed points, with a predetermined start and finish. This distinction both suggests how ashes eventually find themselves in a particular place or geographical location, and describes the trajectories of thought, feeling and conversation which intertwine with these spatial movements. While individuals who 'travel' between *fixed* points – for example, between home and daily work – might feel impatient at being liminally 'betwixt and between' (Turner, 1969), those who walk an open-ended line, for example, wandering through a city as a tourist, participate in its active nature and can develop an intimate relationship with that pathway. This distinction suggests how the route taken by people who retain ashes, for a time at least, and then move them somewhere else, create a movement that resembles the line that goes for a walk. Desiring a meaningful final disposal, interviewees set out without the authority of set liturgies or institutional structures to discover rather than simply arrive at the end points of lines they are establishing. Those who follow a once well-trodden path between the crematorium and the Garden of Remembrance may well be on the other kind of line, with a predetermined endpoint. One interviewee recalled a previous disposal experience, saying, 'It was virtually taken as read' that her relative would go to Garden of Remembrance. Of those who follow this kind of closed line, Ingold says:

> The predicament of people in modern metropolitan societies is the extent to which they are compelled to inhabit an environment which has been planned and built expressly for the purposes of occupation. The architecture and public spaces of the built environment enclose and contain: its roads and highways connect.
>
> (2007: 102)

When informants described the cemetery, then, it was as an institutional site designed and set up in ways that they felt no affinity with; for example, if no one they knew had been disposed of there or indeed visited one of its graves, they might have difficulty engaging with it and so coming to find it meaningful.

Thus, for some interviewees, scattering in the planned environment of the Garden of Remembrance and the cemetery beyond was '*a way of abandonment*', a location where a relative's remains would be '*thrown around*'. Talking about his mother's ashes, one man elaborated on this point:

> *I wouldn't have wanted to leave her there. No, that's a way of abandonment as far as I'm concerned. I wouldn't have been happy with it. Because it's a personal thing in that respect. Again – it depends how you feel about the person. Someone you didn't really care about, you didn't like... it's like... do what you like.*

The cemetery for him was a place that carried no associations, other than death. One person said, '*I just can't associate him there... too remote*'. Another felt she could not leave her husband's ashes at the cemetery because '*he didn't know anyone there*' – and castigated her neighbour for disposing of her own husband's ashes there whilst giving precedence to her cat, whose ashes stood on the mantelpiece.

What Ingold calls 'the structures that confine, channel and contain' (2007: 102–3), his closed lines, are mutable however. For example, through the actions and choices of individuals and groups, well-established death ritual such as the official scattering of ashes in Gardens of Remembrance can take new forms and different trajectories. Thus, ashes may be removed from the crematorium and then returned to a different Garden of Remembrance where the individual has some connection. As Ingold describes, closed lines such as that between cremator and Garden of Remembrance are likely to be 'ceaselessly eroded by the tactical manoeuvring of inhabitants whose wandering lines or efficacious meanderings undercut the strategic designs of society's master builders' (2007: 103). Certainly, the terms 'wandering lines' and 'efficacious meanderings', borrowed from de Certeau (cited in Ingold, 2007: 103), do throw light on the paths taken by people who retain ashes rather than disposing of them in Gardens of Remembrance adjacent to the crematoria that generated them. For such individuals, place-making often becomes a passage to be negotiated between themselves, friends and family, and the deceased person, with the aim that all parties shall co-habit a future that will be recognisable either side of the life/death divide because it fits with their particular and shared pasts. Unlike the smaller proportion of people for whom final disposal is simply a matter of following the established path, or closed line, those who remove ashes after cremation can be the first step along an open line towards an as yet to be decided destination.

Interviewees' accounts of place-making can therefore be seen as outcomes of their being-in-the-world, rather than sets of abstracted comments *upon* past events. Ingold (2007: 90–91) expands his discussion of lines to encompass storytelling, arguing that stories are like paths that are walked, without clear beginnings and endings, but which 'relate' one terrain of lived experience with another. What our data describe is the telling of ash disposal stories that reach back into past lives, family histories and places made years previously. In many cases they do not arrive at a fixed point or conclusion that might contain or limit connections between the living and the dead, the past and the present. As Hetherington says of that other kind of disposal – the dumping of rubbish – the presence of what is *not* can be powerful: 'the erasure of an object is never complete. There is always a trace effect that is passed on by its absence' (2004: 168).

Wandering lines and cul de sacs

If place-making has a participative, negotiated nature, then the influence of the deceased person needs to be recognised alongside that of bereaved individuals, and their associates. Yet who, what and where the dead are is likely to be, at best, fluid. Their opaque status is evoked in van Gennep's ([1909] 1960) account of the ambiguous location of the spirit between first and second burials, one evidenced in much ethnographic work on mortuary rituals (see Robben, 2004; Prendergast, 2005). Similarly, in our data, the ontological status of the dead could trouble those making a place for their ashes and, for example, the dead may not necessarily be confined to a fixed location. A Glasgow interviewee imagined his father only sometimes 'present' with his scattered ashes, now located after a two-year interval at a dog-track in the father's home city of Dublin, a place where he had experienced many good times, sometimes with his son:

> *I think of dad as pretty nearby. I don't think of him as there (at the dog-track). He can be there if he wants – I don't think he's there all the time.*

The suggestion that his father can be present, be 'close by', from time to time, with the son in his Glasgow home, was echoed in other interviewees' accounts. In this case, the man viewed his dead father as able to choose how to conduct himself, as itinerantly independent and moving around, able to be present at different times with people and in places that were his 'life-world'. Another interviewee distributed her father's ashes, in tiny portions, in the pubs he frequented when alive, rubbing them into carpets and upholstery – an example of covert ritualisation (Seremetakis, 1991). She saw these as anchor points for her father's wandering spirit, a route that was as fitting in death as in life. When the son from Glasgow described the dog-track as somewhere his father '*can be if he wants*', we glimpse the challenges involved in place-making that ensures options for the dead to conduct themselves as only they can judge appropriate in a state of being dead that is still uncharted terrain in a post-religious, secular afterlife.

These interpretations are supported by data from a grandson who expressed similar ideas about his dead grandmother. A family decision had evolved over time to scatter her ashes on a hillside with a sight-line to her former house, now owned by strangers, but where the grandfather's ashes had been scattered years earlier. Speaking of this hillside disposal site, he reflected:

> *We went there a lot as a family, a nice place to be, but obviously, she could overlook my grand-dad and then at least, they could see each other – if they actually still wanted to see each other.*

Imagining what it may be like to be dead, and transposing everyday dispositions and desires – to sometimes be present, sometimes absent, from his grandparent's lives to their after-lives – he seemed, momentarily, to find a way of placing himself in their new mode of being (dead). The disposal site linked to the grandparents' habitual places, their ways of being with each other and the surviving family's present, opened up pathways that reached back into time, connecting family members, taking them all along a route into some future place and mode of being.

The same holds for the Dublin dog-track disposal. Accomplished with little fuss, it had nonetheless taken two years for the right conditions to converge, especially for the appropriate time to arrive. The ashes had been in the son's spare room; '*not displayed, just kept... no problem... the ashes are nothing much... but the memories are there*'. Eventually, he contacted the race-track where, he said, staff were initially '*taken aback*' but then helpful. He and his wife were the only family at this final deposition, the others, along with friends, having attended the cremation two years before. On that occasion a priest officiated, but now the offer of priestly assistance was not taken up, suggesting the personal and secular nature of final disposals. All this was represented as highly satisfying:

> *It was nice. I felt great – we completed it... we went to a pub he used to drink in, it had changed names but I knew where it was... we went for a meal. It was lovely.*

The sense of achievement, of something settled, is palpable here. A particular post-mortem intimacy is claimed in finding the pub that no one else could have located and this suggests a shared, even secret, territory. He summed up his efforts and achievement saying:

> *It helps to be able to do something someone has asked.... I'd rather what he wanted than what I wanted.*

The right and satisfying thing therefore seems to entail working out what the deceased person would have wanted to happen to their ash remains once they were fully established in death, not immediately after the cremation.

These two survivors' accounts, the son's and grandson's, reveal how over time they had arrived at a way of 'tracking' the state of being dead via an open-ended process of line-making that involved refractions of day-to-day death grounded in life experiences. Where would the father or grandparent be? What would they be choosing to do at any given time? Who would they need to be seeing or, perhaps, avoiding? In this way scope to exercise future options seems to be built into the way the ashes are dealt with, for both the survivors and the deceased in equal measure.

Interpreting our material in this way reveals the frequency of an unspoken deal or contract between the deceased and the survivor. If references to a place and manner of ash disposal had been made by the deceased individual, these were usually as part of family 'joking': and although only in passing, they were frequent enough to be registered as significant – as with the dog-track. Whilst joking remarks might seem a fragile basis for important, future decision-making – and the living frequently articulated the deceased person's wishes as central to the process – no hint of criticism or complaint about this vagueness was voiced by any interviewees. Instead, disposal options were worked out on the ground and when the time came, these barely formulated plans were edged towards, in conjunction with the remembered or conjectured wishes of the deceased.

Drawing on the notion of the line that goes for a walk, then, it becomes clear that the movement between cremation and later disposal is often a creative rather than predetermined process. It is also participative in that new factors will intervene during the time spent making a decision and these have to be negotiated. To an outsider, events and dispositions can appear one dimensional – bereaved people are apparently making straightforward choices, opting for a closed line between predetermined points. Yet interviewees' hesitant comments suggest otherwise. Even when options to vary the relationship between dead body and an identifiable disposal site seem very limited, some bereaved people may quietly continue to make things up as they move along, experimenting with memorialising practices that *might* 'fit'. Achieving the kind of balance between what the deceased person would have wanted, or said they wanted, and what the survivor(s) were able – practically and emotionally – to accomplish could therefore take time to work out and enact.

Such conditions formed the environment within which place was being made, the being-in-the-world which provided the homeland for interviewees' decisions. Among them were experiences that did seem to close down options, narrowing the scope of the survivor to a predetermined path or line. Many surviving partners simply wished to be reunited with their deceased spouse, the means and locus being secondary, often the outcome of a joint decision. For example, an interviewee who was unwavering in her choice had made provision that her ashes would be placed with those of her husband, in an ash grave at the local cemetery.

The ontological status of the dead and the symbolic efficacy of their partial allusions to the place and manner of disposal were often integral to the line trodden by a survivor, something visited and revisited along the way. That said, where the deceased person had made remarks such as '*It doesn't matter, I don't care after I am gone. You can throw me away... put me in a bin bag*', their remarks had less potency and were invariably circumvented. One interviewee whose father had said he did not care what happened to his ashes, '*Just throw me on the river*' (where he had worked all his life), said that she could not bring herself do that, though she claimed to be unconcerned about the fate of the body. In consequence, some years on, her father's ashes – along with those of her mother who died a year later – had been transferred to expensive ceramic urns and remained in her living room. She explained that an initial hesitation about her father's instruction, which she had taken seriously enough before he died, had then led to indecision and at the time of the interview had culminated in a state where she was almost afraid to act. She felt, however, that given time to turn matters over in her mind and the opportunity to talk about the impasse, as in the interview, she would be able to take action. With both her parents' sets of remains ever present, a resolution would eventually present itself. When this happened she would know it was the right thing to do: '*I'll know when I feel comfortable*'. To be resisted, in her view, was action that she would come to regret. So this interviewee had yet to find the right fit between her father's past and the meaning of his life, one that would mesh with her own life and history.

Those who pre-decease survivors can only *hope* that whatever transpires will be fitting, but a glimpse into the way that 'deals' might be made before a death is provided by interviewees' responses when asked about their own disposal. Most had something to say on this, but only a quarter – predominantly the surviving spouses already mentioned – had made fairly definite arrangements about burial or cremation and where disposal should be. However, the majority admitted to having approached this only obliquely or in joke, when a specific place may or may not have been named. Sometimes a generic location – a hillside, '*a beautiful place*' – had been lightly dropped into conversations about disposal. One informant may be summing up what is going on in these exchanges when he said:

> *I'm open –minded – whatever my partner (at the time) decides. I would make some specifics but leave some scope as to how he finally releases me. I wouldn't want to be just left at the crematorium. It would be nice to be scattered in a place precious to me.*

In contrast to the sense of contentment that characterised these interviewees' comments, a fitting resolution had eluded others. One woman's husband's ashes had, she claimed, been scattered at the crematorium when she had wanted to remove them for scattering on a local lake. This had left her disorientated, without a focus to relate to:

> *What have I done to you – you're just a number now and nothing left of you... It still bothers me that I don't have a memory of where his ashes are... I would have been there, saw the last of his remains... I've never been there. There's nothing sacred there.*

Despite the fact that her husband's ashes were in the crematorium grounds, and the co-ordinates marking an approximate location were undoubtedly available, the interviewee appears to be describing an insurmountable obstacle to connecting with him in his state of being dead. It seems that a block or chasm has interrupted the thread, or line, connecting their past life and any future habitus that might offer shared ground.

Conclusion

Removing ashes for later disposal was thus a choice which set many bereaved people upon an open line of indeterminate direction or length. Its shape emerged out of their being-in-the-world, rather than any free-floating mental representations. It unfolded over a period of time, shaped by the transitions undergone by the dead as their ontological status was scrutinised, if not resolved. In some respects the lines engendered through these processes conform to what Ingold (2007) refers to as ghost lines; that is, lines which are imperceptible yet known and experienced, for example, the Australian aboriginal population's song lines which criss-cross their continent, or the meridian lines that conduct vital forces around the body which the practitioner of traditional Chinese medicine works with. Vital and alive in the accounts of our informants, these lines often trace invisible aspects of their emotional geographies and indeed their environments of memory.

Note

1. We are grateful to the ESRC for their support for the study, 'Environments of Memory: Changing Rituals of Mourning and their personal, social and emotional implications' (2003–2005).

10
Natural Burial: The De-materialising of Death?

Andy Clayden, Jenny Hockey and Mark Powell

Defining the natural burial ground

This chapter describes a research activity designed to explore public understandings of the concept of natural burial and asks how people might make sense of embodied experiences of the materialities of the natural burial ground. As both a concept and a practice, natural burial is particularly prevalent within the United Kingdom with around 250 sites now in existence, compared with 25 in Germany, 13 in the United States and 2 in the Netherlands.[1] That said, uptake of this burial option represents a small proportion of the United Kingdom's overall disposal choices and what the public and indeed providers understand by the term 'natural burial' remains unclear. The data presented here enable us to address this question as part of an ESRC-funded project[2] which explores the cultural, social and emotional implications of natural burial and explores its potential links with the professionalisation of death and dying, a revival of Romantic values, ecological concerns and the claiming of a distinctive identity through bespoke disposal. As the data indicate, answers to these questions are by no means mutually exclusive.

Core to the development of this work has been the diversity of design, management and ownership of natural burial grounds since their inception in 1993. Potentially evident within this diversity are possible orientations towards this practice, reflected in the project's theoretical concerns listed above. In part, this contributes to the difficulty of enabling individuals to make informed choices about pursuing a natural burial option, an issue addressed in the research activity discussed in this chapter. While each natural burial ground may subscribe to a shared aim of providing an environmentally sensitive approach to the care of deceased people and the landscape, how this is addressed may vary significantly (see Thompson, 2002; Clayden, 2003, 2004). This variety partly reflects site providers different habitat objectives; for example, the creation or preservation of woodland or the return of improved pasture to a wildflower meadow. Where the concept of natural burial is expressed in the creation of a new habitat, this may not be realised for many years and, therefore, makes the challenge of communicating the idea even greater. For example, natural burial grounds are frequently described as woodland burial even though no woodland may yet exist. This requires the family and friends of the person who has died to imagine a landscape they may never experience in their own lifetime. It may also require them at some future point when the trees have become established, to accept that they no longer have access to the grave.

For this project, alongside data from members of the public discussed in this chapter, we carried out site visits and interviews with 20 burial ground owners and managers across the United Kingdom. These reveal a richness of design and interpretation of natural burial which goes beyond a purely ecological or environmental aspiration. Perhaps inadvertently, the natural burial movement has opened the door to new operators including charities and not-for-profit groups, natural burial companies and landowners whose motivation for providing burial space may be very different from the Local Authority or Church. These providers bring with them different motivations and relationships with the landscape which may not be informed or confined by a working knowledge of the cemetery or professional experience of caring for deceased and bereaved people. For example, an upland farmer we interviewed, in the North of England, whose family had worked the land for many generations, decided to use one of his fields for natural burial to diversify the farm's income. His approach to setting out the field for burial was informed by his knowledge of traditional farming techniques. By ploughing, he decided to return the field to the traditional rig and furrow, once a common feature of upland farms.[3] This pattern became a template for locating each grave; on the rig, between each pair of furrows there is space for two adjacent graves.

What's new?

In discussing natural burial grounds we begin by exploring their relationship with more traditional burial sites. In defining a burial ground Rugg (2000) highlights the explicit order of the cemetery landscape, where typically each grave is easily locatable within a grid of rows and sections. Within this grid it is commonplace for the name of

Figure 10.1 In the natural burial ground at Usk Castle Chase, South Wales, no grave markers are permitted.

each occupant and their date of birth and death to be recorded on a headstone. Whilst many natural burial grounds permit stone or wooden plaques to be placed flat on the grave, often in addition to a tree, there are sites with no explicit order of the graves and where no memorialisation is permitted. Figure 10.1 shows an open field into which there have been a number of burials. In this model of natural burial only the burial ground manager, with the aid of the site plan, burial register and discrete, yet fixed markers can locate the precise position and orientation of the grave.

However, it is noteworthy that even within this regulated landscape, some bereaved people succeed in capturing and preserving the identity of the grave, as Figure 10.2 beautifully illustrates: a thin strip of grass has been painstakingly clipped around the perimeter of the grave, and two small twigs have been arranged to form a cross at its head. While evident within many natural burial grounds, the unmarked grave is by no means unique to natural burial where the motivation to remove or downplay the status of headstones can be attributed partly to cemetery managers' difficulty in maintaining existing memorials and managing the cemetery landscape (Dunk and Rugg, 1994; Worpole, 1997). That said, from their inception in the early nineteenth century, cemeteries themselves have always contained significant areas or rows of public graves where people who died in poverty are buried, and where their relatives are

Figure 10.2 The desire to retain and mark the identity of the grave is shown here where the grass has been carefully clipped around the perimeter of the grave.

not permitted to erect a memorial. A number of local authorities have now returned to these areas, reclaiming unused burial space in order to provide a natural burial option.

The cemetery landscape is also made legible and distinct by its buildings and infrastructure: a chapel, entrance lodge, maintenance buildings, a perimeter wall and railings, typically locked at sunset (Rugg, 2000). While these features are less likely to typify a natural burial ground, all Local Authority natural burial sites are located either within this familiar landscape or in adjacent land reserved for future burial. Therefore, their physical and administrative context differs from many of the privately owned sites, usually located beyond the urban fringe, within a rural landscape. Unlike the traditional cemetery, access to these burial grounds is not restricted after sunset. The perimeter boundary is also frequently indistinguishable from the adjacent fields and may be formed from stone and earth banks or hedgerows of hawthorn and field maple. Many of the smaller private natural burial grounds do not include buildings for staff or visitors. If visitors should require information about the burial ground or for example they need assistance in locating a grave that is not a directly visible part of the landscape, they are typically directed to the nearby farmhouse or landowner's home. However, not all natural burial grounds conform to this model.

The diversity seen in the range of habitats that natural burial potentially preserves or creates is also reflected in the different levels to which owners are choosing to invest in natural burial. There is a small but growing trend to develop larger natural burial grounds comparable to, and even surpassing, municipal cemeteries in terms of size, the facilities offered and the fees charged. Whilst most natural burial grounds operate a single fixed-price policy, irrespective of the grave location, some of the larger burial grounds vary the price depending on the quality of the setting; for example, a desirable view or close proximity to a mature tree. Differential pricing has also been a feature of private and municipal cemeteries where prominent positions within the cemetery, for example, on the main avenue or near to the chapel, commanded the highest fees. Within the natural burial ground the *visual* prominence of the grave is not a consideration for there is no significant marker. Instead, it appears that (where costs differ) it is in relation to the perceived degree of 'naturalness' and beauty available. While these natural burial grounds may share similarities with private and municipal cemeteries in terms of varying fees, the absence of headstones, plus a distinctive building aesthetic which frequently draws on vernacular traditions such as timber framing, to create barn-like buildings or shelters, sets them apart. However, while these developments contribute to the diversity which has inspired this project, they are not typical of the majority of private natural burial grounds which are far less segregated from the surrounding landscape, both visually and physically. It was within this 'new' burial landscape that we wanted our participants to engage.

Exploring public understanding of natural burial?

As we detail below, we approached this question by taking people with no association with a natural burial ground into a site at Ulley, in countryside between the major cities of Rotherham and Sheffield and within minutes of a major motorway. Their responses to this environment showed the extent to which natural burial was a familiar concept to them, and whether the site conformed to any expectations they might have; in addition these data illuminated the cultural imaginary through which notions such as 'nature' and 'freedom' might acquire meaning and resonance. These terms can be seen as examples of what Cohen (1985) calls 'hurrah' words, terms which elicit an uncritically positive response, yet retain a highly ambiguous meaning. As a result, they can bind potential users of a site to an apparently shared, yet unspoken, set of values.

We took a sample of 25 people into a natural burial ground (landscape) where particular values were espoused and explored the extent to which these terms figured within their responses and, if so, whether they reflected a consistent set of meanings. This method allowed us to examine the relationship between knowledge derived from external sources, such as the media, and embodied experiences of landscapes through entry into them. As Bender argues, the dominant western 'view' of landscapes is surface-oriented and ego-centred: 'a perspectival landscape, a landscape of views and vistas' (1993: 1). Indeed the term 'landscape' is historically as well as culturally distinctive, appearing in the West for the first time during the seventeenth and eighteenth centuries. That said, Bender cites the co-existence of other ways of relating to the land that were concerned with spatial or temporal experiences 'within' landscapes. More recently, Ingold has critiqued the anthropological notion that if social groups and individuals construct different 'readings' of the same environment this simply reflects their scope for imposing their own cognitive schema upon an external world. In his view, this rests upon precisely the western ego-centred, perspectival orientation towards the landscape that Bender (1993) refers to: 'a disengagement of the observer from the world' (2000: 15). In its place, Ingold substitutes 'the whole-organism-in-its-environment' (2000: 19), a totality that he views as a developmental system.

This theoretical stance is complimented by Milton's (2002) argument that an understanding of the world is shaped not simply by culture, but also by the emotional experiences a landscape evokes. While our participants did not have the same emotional connection to the natural burial ground as those who have buried a friend or relative there, they still had an embodied experience of this setting, one referred to by Ingold (2000: 153) as a 'dwelling' perspective. In refuting the notion that thought differentiates human beings from the natural world, Ingold (2000) describes people as simply another element within a shared landscape, their responses deriving precisely from engagement with it. This theoretical orientation informed our decision to take people into a natural burial ground and ask them to explore it firsthand. It is this orientation which steered us away from simply presenting them with photos of different natural burial grounds as visual stimuli for focus group discussion.

Into the field

To access the responses of people with varied experiences of and orientations towards death, we recruited a sample from across the life course – young adults, people in midlife, people over 60 – and worked with them in three separate groups. After preliminary interviews that explored their background, knowledge and perceptions of death and disposal, each group was invited to visit 'a burial ground' with the research team. Many of them were entering a natural burial ground for the first time.

We described the destination of their visit as simply a burial ground to prevent participants acquiring knowledge of natural burial they would not otherwise have had, for example, through Internet searches. However, before they got into the mini bus, we revealed the destination in case they wished to withdraw at that point. The half-hour journey was an opportunity to discuss with one another what a natural burial ground would look like: a process that entwined ideas gleaned from one another with their own experiences and perceptions of 'natural', 'woodland' and 'burial ground' (see also Bender and Winer, 2001).

On arrival, group members divided into pairs. The pairings were quite informal – for example, husband and wife or friends, since the aim was simply to enable the participants to explore the site with someone they felt comfortable talking to and to discuss and record their reactions to the natural burial ground. Each pair of participants was encouraged to walk around the site (see Figures 10.3 and 10.4) and consider the following tasks and questions:

- Describe what you see.
- Is what you see different to your expectations of a burial ground/cemetery, and if so how is it different?
- What do you like and/or dislike about this place?
- In what ways do you think this burial ground is 'natural' or 'unnatural'?
- How do you feel about the way different graves have been marked?

As the layout of the burials, the management of vegetation, and the marking and memorialising practices at Ulley were largely unregulated at this time, participants had to explore the site carefully to address these questions. There was and is, however, little to guide this kind of scrutiny, other than a grass pathway mown by the farmer employed to manage the site. Participants had to negotiate undergrowth and navigate memorial trees and other items placed in the grass, rather than simply walk between obvious vantage points. Therefore, they were obliged to enter into the natural burial ground landscape, undertaking a journey of discovery where their expectations of what a natural burial ground might look like were confirmed or challenged through direct experience. As Ingold explains, such processes of 'thinking, perceiving, remembering and learning have to be studied within the ecological context of people's

Figures 10.3 and 10.4 These two images show our research participants, in their pairs, exploring the burial ground and venturing off the path to investigate individual graves.

interrelations with their environments' (2000: 171). We found that our participants interpreted similar sensory experiences in different ways, depending on how they attuned to this environment. Again, Ingold describes this kind of activity as, 'a process wherein both persons, as knowledgeable social agents, and the settings in which they act, continually come into being, each in relation to the other' (Ingold, 2000: 162).

After each site visit, the pairs of participants were asked to team up with another pair of their choice to form a focus group back at the university. The focus groups provided an opportunity for the participants to reflect on their experiences and to discuss their response to the natural burial ground.

The data gathered during the three stages of this research activity – interview, burial ground visit, and subsequent focus group – give insight into participants' first impressions, the aspects of this burial environment which appealed to or disturbed them. Importantly, they show how participants may have redefined their understandings as a result of their experience. In making sense of these data, then, we considered how participants' prior experiences of death and its landscapes, as identified in the preliminary visit interview, might influence – or be influenced by – what they experienced at Ulley. However, in addition, we located this process within its wider historical and social context. As we go on to show, many of the contradictions evident in participants' discussions reflect a legacy of conflicting ideas encompassed within Romanticism that link with the notion of a perspectival landscape of vistas and views (Bender, 1993). Macnaghten and Urry (1998), for example, discuss the notion that the countryside, as an object of the gaze, particularly a 'romantic' gaze, is experienced by a privatised individual who consumes an 'unspoilt' English countryside. This partial view excludes the presence or activities of other people, and can be linked with tensions between an Enlightenment tradition which saw nature as a site requiring human intervention and a Romantic critique of nineteenth-century *outcomes* of human intervention – urbanisation and industrialisation. Tellingly, in terms of our data, Macnaghten and Urry argue that 'while these many negative impacts were relatively easy to identify; it was much harder to image and to articulate a coherent "natural" alternative' (1998: 12). Therefore, Romanticism as a set of ideas, not only separates the individual from unspoilt landscapes which provide personal inspiration and improvement; it also celebrates the individual's uniqueness, and particularly their capacity for emotional expression (Rugg, 1999). As the data below indicate, these dimensions of a single movement can become opposed within the landscape of the natural burial ground.

Nothing natural?

Given the centrality of contemporary ecological considerations to the inception of natural burial we wanted to know what might make this site 'natural' – or indeed 'unnatural', for people. Did participants have a notion of the 'natural' that reflected Romantic assumptions achievable only if evidence of human involvement had been obliterated: for example, grave marking, gardening or memorialisation? Does an authentically 'natural' landscape require human beings to be somehow detached from 'nature'?

Alternatively, did participants have a more individualistic model of 'human' nature which also reflected Romantic ideals, but was concerned with the emotionality of individuals who, 'by nature', needed to place cut flowers, plastic ornaments or football shirts on relatives' or friends' graves in order to mark out their unique identities – even if these items were somehow 'unnatural'? As one participant acknowledged, '*The stuff on trees, it's not natural but individuals need individual things*'. From this perspective, is such 'stuff' any less a part of a natural burial ground than trees, 'native' plant life and other organic items? Should concepts such as 'woodland' and 'natural burial' be recognised as the outcome of culturally specific activities? Following Ingold's (2000) refusal of any separation of human beings from the environments within which they 'dwell', it is possible view to the act of burial itself as cultural, in that historically, human bodies have been disposed of through a vast range of mechanisms. 'Natural' burial is therefore another cultural activity, its descriptor acting simply to differentiate it from other western burial practices. In the arrangement of organic and 'natural' objects on a grave site, for example, ideas about 'nature' interweave with materially grounded, cultural practice.

First impressions

Data gathered during participants' visits to Ulley reveal them evaluating their experiences in relation to their knowledge of other woodlands and other burial grounds. What sense did they make of a setting where the two combine? For some the site was appealing because, as one participant explained, '*it's like woodland*', '*it's free*' and '*the only thing neat is the path*'. As a burial ground, however, it was felt by others to lack dates on graves, so obscuring the 'story' of the deceased person. Non-organic memorialisation, such as plastic flowers and toys, was not necessarily seen to undermine Ulley's 'woodland' status, and was welcomed by participants who saw these artefacts telling another kind of story about the person who had died. Indeed, those participants who welcomed memorialisation often saw it as evidence of a detachment from '*Catholic Gods*' which allowed people to bring '*their own things*' for memorialisation purposes. This notion of freedom pervaded our data, often expressed in terms of escaping the authority of 'the church', a stance which echoes the aspirations of those who set up largely unconsecrated cemetery landscapes in the early nineteenth century, many of whom were non-conformist (Rugg, 1999: 220).

Other participants, Pat and Dave, regarded the burial ground as a '*contrived*' landscape which contravened the ideal of an '*unspoilt*' English countryside by bearing the marks of human intervention (Macnaghten and Urry, 1998). Therefore, for them the non-native tree species made it an '*un-natural*' woodland that was further undermined by the '*stuff*' and '*tat*' of '*naff*' memorialisation. In this sense, Ulley failed their expectations of both woodland and a burial ground. It could not be likened to '*a tree on the crags*', which for them represented '*a natural, nature made landscape*', and as such, more of a '*natural grave yard*'; it also lacked the '*laid out, neat and tidy*' atmosphere of the cemetery.[4] Reflecting their Quaker background and its commitment to simplicity, expressed in half size, uniform gravestones (Arnold, 2006), both Pat and Dave felt it unnecessary to mark the grave with anything more than a tree. However, a Romantic legacy of privileging the uniqueness and emotionality of the individual is evident in their acceptance of the idea that '*individuals*' have alternative ways of thinking about death and the person who has died, and so bring different needs to the burial ground.

Pat and Dave's contradictory responses to what they perceived as an ambiguous environment took another form for May and Paula, who appeared unconcerned by the planting of non-native tree species and saw Ulley becoming a '*proper wood*' if not a '*forest*' in time. However, while they liked the plantings around the graves, it did make them into individual private gardens which, for them, prompted uneasy feelings of entering private rather than collective space. Their sense of making illicit entry was revealed in their whispers in the vicinity of certain grave plots. The ambiguity of this landscape was also evident in their talk about visiting Ulley for a picnic as they moved into other parts of the burial ground. These contradictory, embodied experiences of the site encompassed aural as well as visual stimuli. Among others, Steven and Eileen enjoyed the sound of bird song, yet disliked the tinkling of wind chimes in the trees.

Ulley therefore challenged many participants in terms of its ambiguity, which drew on contradictory elements within a cultural imaginary that reflects the legacy of Romanticism. Thus many participants revisited their initial assessment of the burial ground, along with their definitions of concepts like '*natural*' or '*woodland*' or '*burial ground*'. For example, Steven and Eileen began by questioning whether a '*natural*' woodland could be created in a burial ground *'littered'* with memorial *'bric-a-brac'*; but later agreed that they both could *'breathe'* at Ulley, that it offered something different from a *'traditional' 'regimented'* and *'claustrophobic'* cemetery landscape. Whilst many, like Steven and Eileen, initially regarded the choice of non-native tree varieties as problematic aspects of *'natural'* woodland, they later decided that *'woodland'* or natural burial was not just about creating *'native'* woodland, it was also about satisfying the emotional needs of individual users.

Digging deeper

Moving from these first impressions of participants' responses to the more detailed indicators of human intervention in this landscape, we consider surface-level memorialisation and the site's below-ground occupants. In this landscape it can be difficult to sustain a western focus on surface rather than depth, since the site's signboard ('South Yorkshire Woodland Burial Ground') indicates the presence of corpses decomposing underground. Grave markings similarly evoke the remains of those individuals whose identities they express, even if to some extent they drew participants' attention away from the materiality of the corpse and towards survivors and the deceased person's previous life. Earth dug up and left on the surface to backfill a grave was, however, a less easily assimilated indicator of what lay under the grass participants were standing on. That said, without the cemetery's kerb sets, the positioning of corpses was only sometimes indicated, perhaps by grass mown by family and friends, or low DIY fencing. Some participants were therefore unsettled by the possibility of standing on a *'grave'*.

When actively 'foraging' within the site, what they found was evidence of how others engage with this landscape: the owner of the site, its manager, the farmer who maintains it, local people who might walk in through the unlocked gate, visitors to the graves and the dead person who had pre-purchased their plot. Therefore, the data document two overlapping processes. Participants were responding to what they uncovered as they moved through the landscape, whilst simultaneously interpreting the emotions, desires, aesthetics and values of other people, as revealed by their residual traces. In stepping into the footsteps of mourners, they were also imagining themselves into a particular memorial landscape. For example, the site is dotted with small benches set up near grave sites, many of them simple three-plank DIY constructions. These stimulated participants to imaginatively identify with family and friends of the dead: for example, Josie, a member of our youngest age cohort, said, *'can you imagine, like, coming here to visit somebody? You've got your little bench you can sit on and it's like being in a little garden, really nice'*. Janet, a member of the midlife cohort, said of one of the benches: *'It was small, so you could sit on top of it, not that I did, but you could sit on top of it and just contemplate and I thought that was quite a nice thing'*. Chloe, a younger participant, imagined herself placing the bench, highlighting the natural burial ground's scope for *'freedom'* which contrasted with the regulated landscape of the cemetery: *'and again there was a lot of people who'd obviously brought them to . . . so they could sit there and again in a normal graveyard you can't do that. You can't bring your own bench and sit down . . . it's a case of, here's the benches, sit there – whereas there it's here's your area, put whatever you want there'*.

As Francis *et al.* showed in their ethnography of major London cemeteries, even in this regulated, professionally managed landscape, friends and family of the dead are far from passive observers who simply pay silent respects at the surface level of a grave. Instead, they argue, 'through the language of stones and flowers . . . burial and the first year's bereavement become marked on the landscape' (2005: 55). Thus, clearing funeral flowers, establishing a floral marker, creating a memorial garden, erecting a stone marker, turfing a grave and gathering family to lay tributes at the first anniversary of a death were among the activities they observed people undertaking. Mourners in the City of London Cemetery and Crematorium, which originally served a white working-class East End community, were committed to tending headstones and grave gardens in the post-1950s lawn section, welcoming its *uniformity*: ' "it's a classless cemetery now", they said, "No one here looks poor or rich" ' (Francis *et al.*, 2005: 49). Although they generally adhered to a shared code of practice and supported management's control of grave-tending deemed excessive, individuals whose practices were curtailed nevertheless would express anger, suggesting a less than homogenous social environment (see Chapter 8). Such tensions are explored in Potts' analysis of criticisms of roadside memorials as 'tasteless and overblown' (2007: 2). She argued that these reveal a fear of both 'inappropriate' emotional expression and mass culture. Thus the disgust expressed towards plastic tributes and cheap, cellophane-wrapped flowers bought at service stations in her view reflects a belief that 'mass manufactured objects constitute inappropriate vehicles for grief' (2007: 4). Edensor and Millington (2009) offer a parallel analysis of expressions of disgust in response to the illumination of the outside of the house with multi-coloured lights at Christmas, one which interprets these critiques as contrasting forms of class expression.

Contradictions inherent within perceptions of 'nature', already discussed, therefore seem to be complicated by tensions within what Edensor and Millington call 'the symbolic economy of class expression and conflict (that) continues to pervade popular culture' (2009: 104). In what they experienced as Ulley's ambiguous landscape, Sandra and Stephanie, for example, members of our oldest cohort of participants, found items on graves *'messy'*, creating a *'tacky'* environment of *'plastic bling'*, *'folderols'* and *'fancy things'* more suited to the burial of a pet than a person. They had expected there might be *'a little tag'* on a memorial tree with the name of the deceased person, but names on stones spoiled the anonymity, or secrecy even, of a burial location known only to friends and family. Other people didn't need to be in on this knowledge. By contrast with their assertion of the importance of privacy, the families interviewed by Edensor and Millington (2009), who illuminated the outside of their houses at Christmas, espoused the values of sociality, neighbourliness and community; Edensor and Millington suggest that 'it is difficult to think of another strategy by which people of limited means could make such an impression on everyday space' (2009: 111–112). While our data do not indicate the class identity of the dead, the symbolic economy that Edensor and Millington (2009) refer to is useful in developing an understanding of significant differences in responses to the marking of a grave. For some participants, then, a proliferation of material items represented a positive opportunity to memorialise. Haley and Josie, for example, members of our youngest cohort, felt that the natural burial ground was a site at which different forms of memorialisation could be integrated: *'instead of having somebody buried in a cemetery with a grave marker on it and then having a bench or a tree planted somewhere else, they've just done it all in the same place'*. So for them the natural burial ground was both a memorial garden and a graveyard, a place where individuals clearly felt free from *'the dictates of the church'* which, in conventional sites, was always *'hovering over you'*, *'breathing down your neck'*.

This freedom to personalise the burial location was appreciated by many participants, although tempered by an expressed need for regulation: *'You can personalise it and your family and friends can personalise it, **within reason obviously**'* (emphasis added). Within focus group discussion among members of our oldest cohort there was mention of the *losses* brought about by freedom from conventional grave-marking with a headstone. Someone felt trees do not in themselves remind living people of their own mortality – and death, as a result, becomes even more hidden. In other words, the distancing or sequestration of death which has arguably resulted from the professionalisation of deathcare (Mellor and Shilling, 1993) can paradoxically return by the back door in that many natural burial ground users make this choice to *avoid* the services of professionals. Moreover, a sense of history of earlier communities, of the collective nature of death, of '*part of the social fabric of society*' and of personal biography were all felt to be jeopardised by digressing from conventional records on headstones. These views contrast with those of individuals such as Janet, a member of our midlife cohort, who said that the memorial items allowed her to connect with the men or women buried there, to '*feel*' who they had been. In addition, they manifest the care of survivors and as such made her happy. Indeed, neglected or wilting memorial items signalled a lack of care for Janet and disturbed her. Like other participants, Janet also found that exploring the memorials gave her a sense of the creativity or even playfulness experienced by bereaved people. Individuals made references to the children's book, *The Secret Garden,* along with comments such as '*there's so many lovely little things at the bottom*' (of a tree). One of youngest cohort recollected a miniature landscape created on one burial site which had '*a little picket fence in, and it was all sort of scale and there was a wooden man sat on it and then he had a bird and . . . the big sunflower going round*'.

Whatever participants' personal responses to memorialisation were, they frequently commented on the need for people to be able to do what was right for them and often ended their discussion with an acknowledgement of the difficulty of ever making hard and fast rules. For example, Terry, a member of our youngest cohort, speculated about possible contradictions experienced by bereaved people who might feel that they '*should*' let their loved one go, quite literally, back to nature. He saw the natural burial ground consuming the dead – and families and friends wanting to hold them back, marking the grave in response to an '*instinct*' or '*natural urge*'. Therefore for him, memorialisation was a forgivable human weakness, the triumph of '*the little trivialities of life*' over the more '*lasting gesture*' of leaving nothing but a decomposing body and a tree behind. The unique individuality of both the dead and the living and their mutual emotional attachment, key features of a nineteenth-century Romantic framing of death (Rugg, 1999), thus appear to persist within the cultural imaginary being drawn upon here.

What these data indicate, therefore, is the dialectic process through which a particular cohort of individuals came to 'know' a particular natural ground, a process which will in some way inform their future disposal choices. They reveal the personal and social resources through which individuals made sense of this setting; for example, their experiences of the cemetery, of the church and consecrated burial grounds, of the countryside. These constitute what Merleau Ponty describes as 'the homeland of our thoughts' (1962: 24, cited in Ingold, 2000: 186), in other words, the influence of previous embodied, materially grounded experience. These resources are brought into the natural burial ground by participants, an environment within which they then engage, through both their senses and their emotional responses to their surroundings. In addition, these data reveal the complex legacies of particular belief systems and their role within contemporary identity-making processes. Developing an understanding of this complexity helps makes sense of public perceptions of a natural burial ground in ways which can effectively contribute to future policy-making around the provision of informed choice. Alongside the landscape itself, the co-presence of both other participants and absent grave visitors' imagined aspirations and emotional needs contributed to their responses. Thus, while they may be aware of elements within the landscape that jar upon them, they not infrequently showed recognition of the importance of the diversity of bereaved people's desires. In other words, their responses were inter-subjective, or social. As such, they reflect Walter's (1994) argument that contemporary responses to death can be described as 'neo-modern'. That is to say, although choice, or the opportunity to 'do it my way', has become a contemporary watchword, as our data demonstrate, 'even when people die in their own way, the death has been constructed together with others' (Walter, 1994: 189). As argued, we might relate to these others in terms of social factors such as class, and via legacies from the cultural history of landscape and burial practice within the United Kingdom.

Taking it further

The research activity described here took place towards the end of summer in order to capture the site when it was still verdant; the grass tall and the trees in leaf. This single snapshot, however, fails to embrace the dynamic nature of this natural burial ground and how its expression changes throughout the seasons. While this seasonality can be imagined, what cannot be seen and experienced is evidence of grave visiting and memorialisation that requires extended dwelling in this landscape. However, regular visits throughout the year have revealed grave marking which were not visible at the time of our participants' visit. They include items familiar in the cemetery, for example, Mother and Father's Day cards, Christmas wreaths and decorations and cut flowers. Regular visits also expose more subtle, ephemeral marking quickly lost with the passage of time. Examples include rose petals perhaps gathered from the domestic garden and sprinkled on the grave and a daisy chain made in quiet contemplation left hanging from a tree. In addition, there are the plants; the memorial tree which through its own pattern of blossom and fruit may encode for the bereaved family dates of conception birth, and death (Clayden and Dixon, 2007), and the annuals and bulbs which may temporarily transform a grave that might otherwise appear lost or abandoned. This study only begins to unravel the complexity of dwelling in one natural burial ground. Indeed, each burial ground will present an entirely unique experience by virtue of its location, habitat, management and the community that dwell within it. Moreover, there will be much more to be learnt by examining these sites in the future, when they are no longer active burial grounds, but have instead become the habitat that was once envisaged, such as woodland, or have reverted to their original agricultural function.

Notes

1. The figure for UK sites is drawn from data held by the Association of Natural Burial Grounds which has been supplemented by data gathered as part of the research project. The majority of UK sites are extensions to existing municipal cemeteries but a growing proportion are privately owned, independent sites which are managed by different landowners including farmers and funeral directors. The figure for the United States was taken from the Centre for Natural Burial website (http://naturalburial.coop/USA/ (10 December 2008) and the figure for Germany from the FreidWald website (http://www.friedw.de/Startseite.AxCMS?ActiveID=1001 9 December 2008)). FriedWald is not strictly comparable to natural burial as it only provides for the burial of ashes in a mature woodland setting.
2. We are grateful to the ESRC for their support with this study (*Back to Nature? The cultural, social and emotional implications of natural burial*).
3. Rig and furrows are more typical of upland areas in the United Kingdom as opposed to the more common ridge and furrow which are associated with lowland areas.
4. Participants had no problem interchanging words like 'burial ground', 'church yard' and 'cemetery'. While academics might have fixed ways of defining each of these as specific types of place (see Rugg, 2000), our participants did not always do likewise.

11
What Will the Neighbours Say? Reactions to Field and Garden Burial

Tony Walter and Clare Gittings

In modern Britain mourning has been largely a private affair. However, since the 1970s the proliferation of public spontaneous shrines and informal memorials challenge mourning's conventional boundaries, making it far more visible in public space. This chapter looks at the taken-for-granted phenomenon of burying the dead in the *public* space of churchyard or cemetery. As the data presented shows, however, very occasionally these take place in a garden or on some other entirely *private* plot of land. It explores reasons why this might not be acceptable – to other family members, to neighbours, subsequent owners of the land, presenting examples of disposal which nonetheless seem to indicate a re-ordering of place, boundaries, public and private.

In academia, the news media and the Internet, there has been considerable debate about shifting boundaries around death, the disposal of the body and sites of memorialisation of the body (Howarth, 2000). What Santino (2006) calls spontaneous shrines, and Walter (2008) the new public mourning, may take various forms such as roadside shrines, memorials on mountain tops, floral tributes to dead celebrities or the proliferation of ritual silences (Grider, 2005). These informal shrines and memorials are hailed by some as bringing death and mourning out into the open, but denigrated by others as inauthentic look-at-me grief, 'grief lite', inappropriate reminders of mortality (O'Hear, 1998; West, 2004). One camp welcomes this celebration of human emotion and its challenge to a supposed death-denying society; the other camp asks not that death be denied, but that it be bounded, so that we can go about our everyday lives without *memento mori* confronting us on every street corner, park bench or mountain summit (Jones, 2005; Mountaineering Council of Scotland, 2008). The discomfort is caused by what had hitherto been bounded (in a cemetery, at a defined war memorial) or private (grief) leaching out into everyday public space. More than uncomfortable, it can be experienced as tasteless, even disgusting. The new public mourning thus comprises matter (memorials, emotions) that, for critics, is out of place (Douglas, 1966; Petersson, 2005). The debate has focussed on grief, which many feel should remain private, invading public arenas, and is similar to debates on the 'permissive' society: some welcome a less repressed approach to sex; others, while not disapproving of sexuality, consider it should remain within the boundary of the marital bedroom and not dominate the media, fashion and advertising.

Burial not in a public cemetery or churchyard, but in your own garden or on other private land breaks conventional boundaries, not by making public what normally is private, but by making intensely private what normally is public. Burial on private land is extremely rare in the United Kingdom, but it is legal (Bradfield, 1993) and of considerable theoretical interest. Does it disturb people because it fails to place the dead in a conventional death place, or is it acceptable because the grave is private, not in the public's face? What does burial on private land tell us about the shifting boundaries of contemporary death (Howarth, 2000), its public and private faces (Mellor, 1993)?

Ever since the Christianisation of Europe, it has been normal for Britain's dead to be buried in churchyards or other Christian burial grounds.[1] From the mid-nineteenth century, cemeteries have supplanted churchyards as the normal place of disposal, augmented in the twentieth century by crematoria. So are garden and field burials matter out of place (Douglas, 1966)? Are they an example of highly charged heterotopia (Foucault, 1984) – a term borrowed from medicine meaning the presence of a tumour in a place where one would not normally be found (Petersson, 2005: 73)? Is a body in your garden abject, an object of horror (Kristeva, 1982)? Are unbounded dead bodies as disturbing as unbounded dying bodies (Lawton, 1998)?

Or, are garden burials part of a culturally accepted romantic Rousseauian tradition? Jean-Jacques Rousseau was initially buried in 1778 (before removal in 1794 to the Panthéon) in a garden tomb on the isle of poplars in the garden at Ermenonville outside Paris. The garden was planned by Rousseau's patron Count Louis-René Girardin, who was inspired by Rousseau's philosophy of the nobility of nature, and is one of the first French gardens designed in the informal English style. Rousseau's island burial, connecting wild nature and grief, tapped into the heart of Romanticism and the new worship of nature (McManners,

1981). The idea of garden burial has earlier origins still in England (Draper, 1967; Thomas, 1983), and we have documented a number of eighteenth-century examples elsewhere (Gittings, 2007; Gittings and Walter, 2009). In 1997, Diana, Princess of Wales, was buried in a lone grave on an island in the Spencer family estate at Althorp and, though many aspects of her funeral drew criticism, none was directed at her final Rousseauian resting place. Relatively few today may know of Rousseau's grave, but that Diana's drew no criticism indicates public acceptance of the concept of garden burial, or at least of the concept of garden burial surrounded by water and an extensive estate, protecting the dead from the living and vice versa.

Whether everyone is so accepting of a body in the suburban garden next door is another matter. We have found that private burial, though embraced by the person arranging it, can sometimes cause offence to others who become aware of its existence. Yet some are not at all offended. We explore the ambiguity of the buried body as potentially vulnerable, in need of protection; and as potentially dangerous to the living, who themselves need to be protected (Hertz, [1907] 1960; van Gennep, 1960). The incorporation of the buried body within nature or the home may feel 'right' to family members, a fitting place where the dead can rest in peace, yet the unmarked, unbounded grave may threaten others who need the dead to be kept in their place, marked off from everyday life. In this chapter, we explore how comfort and discomfort are produced. What makes a private burial site, to use Petersson's phrase (2005), a proper place of death? To address this question we drew on ideas relating to boundaries, and the tension between the vulnerable dead and the dangerous dead, and ask is it the dead, or the living, that boundaries protect? Elsewhere (Gittings and Walter, 2009), we explore the related issue of liminality and examine more closely the difference between bodies buried in the cultivated garden close to the house and those buried on other, usually less cultivated, but still private land.

The study

In 2008, through contacts in the Natural Death Centre, we interviewed five people who had arranged private land burials in Britain in the preceding 15 years. It turned out that they all had buried not in their own garden, but on a piece of marginal agricultural or otherwise uncultivated land, adjacent to their garden or at a distance, owned by themselves or by someone else. Three sites are in southern Scotland, one in northern England, one in southern England. The interviewees were middle or upper class with access to land, but with left/green politics; in age they ranged from their thirties to sixties; four buried spouses who had died in youth or middle age; three have published short articles about the funeral (Speyer, 2001; Johnston, 2004; Hale, 2005). Here we quote extensively from one interviewee who highlighted a number of issues addressed in this chapter. In line with our earlier experience of interviewing people about funerals and memorials (Walter, 1990), our interviewees asked us not to anonymise their dead; we consider that their wishes supersede paternalistic social science research ethics guidelines that require anonymity. Though anonymity may in most social science research be an admirable default position, in research into funeral and post-funeral practices we consider it unethical to anonymise those whom the living seek to memorialise, unless the living ask otherwise (Grinyer, 2002).

Interviews were also conducted with a couple who live in a house which has two previous occupants buried in the garden (Croft, 2007), and with a 'green' funeral director who told us about four home burials – these involved the burial of a son, an uncle, a grandfather and a commune member; those of the son and uncle were in the small gardens of ex-council houses. In addition, there are a number of published accounts of garden burial, for example, Speyer and Wienrich (2003) and *The Telegraph* (2001).

Though burying bodies on private ground in the United Kingdom is rare, privately disposing of dry cremated remains is not (Prendergast *et al.*, 2006). Unlike in Germany and Scandinavia (Petersson, 2004: 46), British families have the right to take ashes from the crematorium for private disposal. In increasing numbers, they bury or scatter them in places of personal significance – burial may be in the garden, scattering is often in symbolic, liminal places such as beaches, mountains or football pitches (see Chapter 9). Below we cite some findings of a recent study (Prendergast *et al.*, 2006; Hockey *et al.*, 2007a, 2007b) of private ash disposal which shed light on our themes of boundaries, protection and privacy. We now turn to the factors that seem to us to be associated with the production of a proper, or improper, place of burial and consequent comfort or discomfort.

Boundaries

Thomas Hollis died in Dorset in 1774 and was buried, according to his biographer, in 'a grave ten feet deep' in a field 'immediately ploughed over that no trace of his burial-place should remain' (Blackburne,

1780: 481). Two centuries later and 400 miles north, Heather Johnston buried her uncle and mother in the field adjoining their cottage in the Scottish borders (Johnston, 2004). She and her husband own the field, one on which a local farmer grazes sheep. Though she has planted a tree and placed an uncarved natural boulder at the head of her mother's grave, Heather has not fenced the graves or otherwise distinguished them from the rest of the field:

> *When I go and walk down the field I can see the trees growing, and when the earth sinks we fill it up with mole hills, and we look after it over the years. Try and make it back to pasture land, with sheep grazing over it. The farmer was a bit shocked about that, 'Oh, you'll be fencing it off?' 'No.' 'Isn't that a bit unseemly?' He was just under 50, and it was funny, because he was very modern, but he just thought this was a bit unseemly, so that was surprising.*

There are five other cottages nearby. Even though the residents of four of them expressed no reservations about the burials, one had concerns, as Heather told us:

> *One of them was really disturbed, thought this was a very strange idea. 'Ooh, ooh, is this right? I'm not sure.' She didn't want to talk about it much, but I think what was on her mind was the idea of having a body so close. Whereas in fact, it was not close. You have to cross the wee road, and then walk, I don't know, a hundred yards or more, down the slope. It's not actually obvious at all. I think what was going on for her was it really made death more obvious, and that disturbed her. She's somebody who pushes away things she doesn't like. Whereas for us, what I liked about it was that it made it more personal and connected and real; those were exactly the bits that disturbed her.*

The farmer's feeling that a fence would be appropriate and the neighbour's subjective sense of the nearness of graves that objectively were further away than a churchyard might be to a house suggest their gut feeling that dead bodies need to have a boundary around them, protecting the living from the dead. Or perhaps they felt the bodies were insufficiently protected and respected. Either way, field burial appears to be more contentious than island burial. Douglas (1966) has argued that matter that crosses key symbolic boundaries may be deemed either sacred or polluting. A grave that to Heather is 'personal and connected and real', a proper place for the dead, and acceptable to four of her neighbours, is for the fifth disturbing and unseemly.

Dead bodies can, in this respect, be like cars, children, parties and washing. I may take great pride in my car or my children, I need to hang out my washing to dry, and my parties are fun. But to neighbours, other people's cars and parties are polluting, their children a nuisance and their washing an eyesore. This is especially likely when these bodies and objects are not controlled, not placed within proper bounds: when someone else's car is left outside my house, when their washing is hung on the wrong day, when I do not know what time their noisy party will end and when the shouts of their children invade my territory. I need to take care of and protect my car, my children and my husband's remains; others, though, may need to be protected from my car, my children and my husband's remains. They are all both vulnerable and potentially dangerous.

It is not just members of the public, but other members of the family who may prefer the grave to be marked, for example, by traditional grave accessories. Heather continues:

> *On the anniversary of her death, or was it Mothers Day, my brother and his wife arrived at my house with one of those manufactured holders for flowers that you see at graves, with the holes for flowers, it says 'In memoriam' or something. Well they arrived with that, and with flowers to go in it... And for him obviously it formalised something really important. What has struck me quite a lot since then is that what feels really good for some people in the family takes a bit of getting used to for others.*

The same divergence is found in some woodland burial grounds where an agreement states that individual graves should not be marked, yet some families nevertheless mark theirs like conventional graves, with little white picket fences, plastic flowers, gnomes and handwritten anniversary cards (see Chapter 10).

Pets and ashes

Though it is extremely rare in Britain for whole human bodies to lie undifferentiated from the natural and/or the human world, this is very common for human ashes and the bodies of pets, both of which may be buried in the back garden or unmarked in some other place. Arguably, ashes and pets are seen as less dangerous, because – in the case of ashes –

they are dry and comparable to bones rather than a fleshy corpse, and – in the case of pets – they are already more part of the natural world.

Many British suburban gardens contain do-it-yourself pet burials, some marked, some unmarked. Though the number of commercial pet cemeteries is on the increase, pet burial in your own garden is not known for causing discomfort, to family or neighbours. Gardens are seen as a place for the natural cycle of growth and decay, which may be why the burial there of non-human species is uncontentious. Insects and mites in profusion also die and decay within the house, but their remains are dry and largely invisible, with discernable remains vacuumed up and removed. Dead mice and birds in the house, however, are out of place, and domestic cats that bring them in may be perceived as naughty.

Ashes are sometimes buried in the back garden. Hockey *et al.* (2007a) give an example. Carol interred her father's ashes under a birdbath in her daughter's quite small garden; her father liked to go and sit in this garden and Carol thought she herself might move house, so her father was more secure in her daughter's garden than her own. Carol 'had cared for her father on a daily basis while he lived in a flat opposite her marital home' (Hockey *et al.*, 2007b) but considered her father's sudden death as bad – alone, not among family – so his final location restored his place at the heart of the family. Though the ash grave is physically marked by the birdbath, visitors would not be aware of this unless told. Symbolically, in death as in life, there is no boundary between him and the family (Hockey *et al.*, 2007b).

The scattering of human ashes to wind or water is of considerable theoretical significance. The mixing of ashes with the elements may be symbolically satisfying for mourners, but problematic for others. At the funeral of folk singer and hill walker Ewan McColl, his widow Peggy Seeger and their children performed one of his songs *The Joy of Living*. The first verse sings of the 'the air like wine' of the northern hills McColl so loved. Having bid farewell in verses two and three to his wife and children, McColl instructs them to take his ashes to 'some high place' and scatter them to the wind so that he may be part of the air they are breathing. This means the absence of any boundaries between McColl's remains and the pure mountain air. Other may love the hills, but remain resistant to the idea of breathing mountain air polluted with bits of McColl, however much they enjoy his songs! But this is the symbolism of scattering ashes to the wind (even if in practice, scattered ashes fall to the ground within a few feet and then get mixed by the rain into a soggy heap). Symbolically, scattering returns us to nature, all boundaries dissolved. Scattering on the ocean has similar symbolism, and physically the ashes actually disperse in accord with the symbolism (Walter, 1994).

The symbolism of scattering to the wind is acceptable because, should any member of the public actually breathe a particle or two of scattered ash, they will not be aware of it. This contrasts with the less acceptable sight of both ashes on the ground and visible smoke from a crematoria chimney. Scattering ash on the wind or in water confirms an obvious, but important, point, that lack of a boundary between human remains and the living has the potential to disturb only if living people are aware of it. What disturbs people is the *perception* that there is no boundary.

Public and private statements

Though we wanted to know only about burials on private land, we could not stop three of our interviewees recounting at length and with pleasure all aspects of the funeral, which – as is common after the deaths of those in their fifties who are likely to fully socially engaged – drew substantial numbers of mourners and were thus very public events. They contrasted markedly in two of these cases with the subsequent privacy of the unmarked grave, whose location would be known only to those who attended the funeral. Visitors to churchyards and cemeteries often read the inscriptions on graves with which they have no personal connection: such inscriptions are both intensely personal and manifestly public. By contrast, unmarked and uninscribed graves (whether or not on private land), like unmarked ash scatterings and private ash burial, are private. (Though in Scotland and in upland areas of England and Wales there is a legal right to roam, not all the field graves we examined in such areas would be recognisable as graves to the casual passer-by). If spontaneous shrines make public what previous generations had kept private, an unmarked grave in a field, like an ash scattering, renders private what is normally public. (Graves in gardens, possibly, may be less private. Gardens typically have more visitors than non-garden plots or fields, and a marked grave in a small garden would be highly visible).

Therefore, is it wise for those who bury on private land to tell the neighbours or other parties? Though this was not an issue for our five main interviewees, there is evidence in some other cases that this has caused trouble retrospectively, and prospectively some householders have worried who to tell that there are bodies in their garden. In the following cases, despite the grave being in a private place, other people experienced discomfort.

> When Terry Lee made the mistake of telling the local authority he intended to bury his wife at the rear of their former council house in Dover, the neighbours were outraged, and their protest led the council to take out an injunction to stop him.
>
> (Telegraph, 2001)

Terry's mistake was to make visible what might, with considerable care, have remained invisible. What was essentially a private statement became public.

In 1996, Barbara Vessey buried her husband in an unmarked grave on the field adjoining their house in Wales, with no objections expressed (to her knowledge) by the local community. Some time later, she sold the land, and 'within weeks, a neighbour who had been frustrated at not knowing exactly what we had done, told the owners a fairy tale....' Not being able to identify the precise grave site, the new owners obtained official permission to hire a mechanical digger to dig up a wide area in order to locate, exhume and cremate the body (Speyer and Wienrich, 2003). Gittings and Walter (2009) provide eighteenth-century examples of exhumation following garden sales.

Heather Johnston – aware of possible future interest not only by subsequent owners but also by family members, genealogists, police and planners – was not entirely happy about the non-recording of field burials in Scotland, and would in principle support a national register. In the meantime, she intends to use GPS technology to prevent any future problems:

> *It is our intention to precisely note the grid reference location of the graves and place this information with the deeds of our house.... The land would be sold with a covenant that the graves should remain in perpetuity.*

She and her family have also had discussions about fixing a marker on the tree indicating this is a grave and of whom. At the entrance to another field burial site, also in Scotland, the family have written a poem on a bronze plaque, saying '*This is where we lived our lives, this is where we gladly died.*'

There is also the question of whether to tell the children. If you buy a house with human remains buried in the garden, do you tell your children? Little Clarendon, dating from 1600, is a modest house with a modest garden in a Wiltshire village, restored in the early twentieth century by George and Mary Engleheart who lived there till their deaths in 1936 and 1948 respectively; Mary willed it to the National Trust.

The garden is occupied by the deceased Mr and Mrs Engleheart, while the house is currently occupied by John and Aileen Croft and their children (Croft, 2007). Aileen told us she did not at first mention the graves to her children, but when she did they had no problems with it. The youngest daughter (aged 6) was fascinated that two of those who lived there are buried there; she points out their gravestones to visitors on the days the property is open to the public. In this instance, time may also be a factor. The Englehearts' remains are presented as history; they presumably have become safe dry bones, no longer in the liminal state of wetly decomposing flesh, more akin to ashes.

While some gardens become a place of burial, some burial grounds become gardens. Anderson (2009) reports on a house recently built within London's historic Highgate Cemetery; the cemetery is effectively the house's garden. In another English city, within half a mile of the home of one of the authors there are eight old (pre-1900) but now full burial grounds. Two have had private houses built on them. A question for the owners is, how public should they make their garden's necropolitan past? In each case, the house name gives a hint: Yorrick (referring to the grave digger in Hamlet) in the case of a former Unitarian burial ground, Eastergate (referring to the resurrection) in the case of a former pauper burial ground. The name does not directly announce the garden's original use, but might prompt a question to the owner about the reason for the name, in which case the previous use becomes known. When clearing the former pauper burial ground, the first owners, who were practising Catholics, wished to respect its dead. They came across many bones, which they placed in a structure resembling a well that they constructed for the purpose; they also commissioned a small statue of an orphan girl, which they placed under a bush near the house. Like the house's name, the meaning of neither well nor statue are immediately obvious to visitors, but may prompt questions that reveal their meaning. When after several years the house was sold, the new owner faced Aileen Croft's question: should she tell the children?

Of course, we might expect these ancient dead, presumably reduced long since to dry bones, to be seen as less dangerous than the recent, fleshy, decomposing dead. They are more akin to ashes. But these owners did worry over whether to tell the children. Though the Englehearts and the original owner of Eastergate were comfortable with the dead at the heart of their ancient garden, subsequent owners were not quite so sure. Feeling protective towards the dead now in their care, they also felt protective towards their living children. As Ariès (1962, 1981) has argued, sensitivities about the dead and about children, historically over recent centuries, have increased in tandem, though this does not mean there is today agreement as to how best to protect children (Pilcher, 1995)

Unmarked graves are uncontentious if not known about by other parties, but can become extremely contentious if their presence is discovered, as we saw with the bulldozing of the Welsh field. Whereas with spontaneous shrines, the concern is that reminders of death can turn up anywhere, with the unmarked but hard to locate grave, it is the dead themselves that could be anywhere (see Chapter 5). And that disturbs those who, not unreasonably, wish death (perhaps along with sex, madness and other challenges to personal control) safely bounded, and it worries parents who wish to protect their children from such dangers.

Conclusion

The creation of boundaries around some of life's more physical aspects – eating, defecation, sex, death and decay – has been extensively analysed and theorised (Douglas, 1966; Elias, 1978; Giddens, 1991). In the United Kingdom, during the nineteenth and twentieth centuries, for example, human remains have been progressively moved away from human habitation to cemeteries and crematoria on the outskirts of town, thus distancing them from the world of the living (see Chapters 8 and 10). Death has been sequestrated (Mellor, 1993), though arguably some of these boundaries are currently dissolving (Howarth, 2000). Mellor and Shilling (1993) have argued that in modernity death is absent in public but very present in private. This has been challenged by Walter (1994) who points to the highly visible presence of death in the media, in medicine and in public statistics; the problem being more that these representations of death are typically dissonant with private experiences of loss. For Petersson (2005), spontaneous shrines in public places can cause discomfort precisely because they bring the abject into the open.

In this chapter we have used the distinctly unusual phenomenon in Britain of field and garden burial to explore these issues further, and our findings align well with the general argument of Douglas (1966) about boundary breaking, and Petersson's (2005) specific argument about spontaneous shrines. The lack of boundaries implicit in private burial can induce comfort or discomfort, a sense of almost sacred fittingness or a significant sense of unease. The potentially most disturbing human remains or other *memento mori* are those that are not only unbounded, but also make a public statement and are highly visible. The current proliferation of spontaneous shrines and informal memorials are good examples. Unbounded graves or memorials cause no trouble, so long as they remain private statements, invisible to strangers. We also conclude that what matters is not so much whether the remains, or other memorial, are in a public *place* as whether they make a public *statement*. Privately scattering ashes on a popular mountain summit is no problem, but can become so if lots of other people do it and lots of others know.

Our analysis has benefited from the fundamental insight of van Gennep ([1909] 1960) and Hertz ([1907] 1960) that the dead can be both *dangerous* to the living and *vulnerable*, in need of protection from the living. Typically, the recent dead are both more dangerous and more vulnerable than the ancient dead. As far as the dangerous dead are concerned, what is lost in burial in the back garden is a boundary between death space and domestic space, a boundary which can contain the threatening dead. Having them in domestic space is rather close for comfort, which may be why those we interviewed chose not to bury in their garden, but in a field. Decomposition in the garden threatens all the kinds of civilising processes that Elias (1978) and Ariès (1981) discuss. The dead need to be bounded and they need to be kept apart from everyday domesticity, or at least that is the normal convention that home buriers are breaking. It is not surprising if the neighbours are disturbed. Field burial locates the dead away from domestic space; but unmarked and unbounded, for some neighbours, this too can seem threatening.

The recent dead are also vulnerable. For some, burial on private land offers the dead protection from a wider public, keeps them close to home or within nature. They may feel the dead are protected better there than in a cemetery. The ashes of the vulnerable dead are often buried in gardens, and Hockey *et al.*'s (2007a) interviewees were very clear that the remains needed protection not provided by the neglected, anonymous cemetery. But others may feel it is precisely the cemetery that protects the dead. Maybe the neighbour who thought the unfenced field burial was 'unseemly' and the relative who brought the flower holder felt that the corpse was insufficiently protected and respected. The hybrid nature of human remains as both threatening and vulnerable makes field and garden burial potentially divisive, and helps explain why individual feelings on the matter differ widely.

Our analysis may shed some light on the rapidly growing popularity and sympathetic media coverage of the now over 250 natural burial sites that have developed in Britain since 1993 (see Chapter 10; Clayden and Dixon, 2007). Natural burial often entails interment in a grave with only minimal marking in a wood or field, sometimes destined to become a wood, replete with romantic Rousseauian nature symbolism adapted to an English love of deciduous trees and small woods (Schama, 1995). But though each of the graves are very unlikely to have any kind of boundary akin to a cemetery's kerb sets, and are therefore symbolically part of nature, the natural burial ground itself – like Rousseau's and Diana's islands – is typically marked out, as a death place, via signage at the entrance. The public is thus protected from offense by inadvertently wandering among dead bodies 'that could be anywhere'.

The historical and contemporary links between the garden and the cemetery have been well documented (Francis *et al.*, 2005); we know that western cemetery design from Père Lachaise onwards has been influenced by garden design and by romantic concepts of nature (Curl, 1993). In this chapter, we have looked at those rare cases, not when a public burial ground resembles a garden or nature, but when nature or a garden actually becomes a site of private burial, and when a public burial ground, bones and all, becomes a private garden. We hope our guided tour of a few of these unusual gardens and fields has illuminated wider issues about boundaries; the historically evolving relation in Britain between the dead, the home, the garden and nature; and the ambiguous status of the buried body as both vulnerable and threatening.

Acknowledgements

We acknowledge helpful comments on an earlier draft by Jenny Hockey, Leonie Kellaher and Malcolm Ramsay.

Note

1. In the United States, the church never had a monopoly on burial. Family burial sites, common in settler days, are still in use in some remote areas even today (Sloane, 1991: 14–17; Amy Whitehead: pers. comm.).

12
Memorialising the Suicide Victim: 'Walking the Walk'

Caroline Simone

> *...but it's a scary walk. It's a really scary walk. And I think, did she get to this corner and think, 'I'm going to do it now?' Did she get to the corner and think, 'If I just step out and if I take one step to the right, then 'smack' and that's it?' Did she think that?*
>
> (Rose)

This chapter foregrounds the experiences of two people bereaved of their parents by suicide, showing how they used the physical environment long after the event of the death to retrace the final hours and moments of their parents' lives. Core to the chapter is evidence that suicide can be so overwhelming that bereaved people choosing to search for meaning and a connection with their dead parent may do so only after many years have elapsed (Shepherd and Barraclough, 1974; Wertheimer, 2001). Drawing from a study on suicide loss experiences, *Bereavement after Parental Suicide: Transcending Chaos and Disorder*, I suggest in this chapter that as a strategy for creating memories, bereaved people may undertake a private 'pilgrimage' – sometimes located in public secular spaces, to retrace their parent's final hours. Thus, one participant, Rose,[1] *'walked the walk'*, retracing her mother's last footsteps 25 years after her suicide. Thirty-eight years after his father's death Will *'re-enacted'* his father's last hours. Prior to their respective pilgrimages, these landscapes had been saturated with negative associations and emotions for Rose and Will. Yet by undertaking what Davidson *et al.* (2005) term an 'emotionally spatialised pilgrimage', both individuals were able to re-evaluate their emotions and more positively assemble their loss experiences (Davidson *et al.*, 2005).

The research that underpins this chapter comes from a project that used an interpretative, phenomenological approach and, whilst underpinned by sociological thought, also drew on literature from psychology, health and counselling. Informed by my roles as a trustee of a national suicide bereavement organisation and a leader for a self-help bereavement group, it also drew on my experience of therapeutic work with practitioners. As such, this was an academic project that sought to enlighten and support people bereaved through parental suicide in terms that related to their lived experiences and professional practice (Bridging Work Group, 2009).

The traumatic nature of parental suicide

One of the most challenging experiences that may occur to a child or young adult is the death of a parent, and evidence suggests that death by suicide further complicates bereavement adaptation (Shepherd and Barraclough, 1974; Raphael, 1984; Pfeffer *et al.*, 1997; Sethi and Bhargava, 2003). For example, suicide may well be sudden and result from self-inflicted violence (Currier *et al.*, 2006), features which can precipitate a tidal wave of emotional and social chaos that is compounded by the likelihood that the cause of death will not be publically acknowledged or socially validated (Corr, 2002). This can have consequences within both the child's family and their wider social networks. Thus, to protect their surviving parent and other family members, bereaved children and young people may often hide their pain (Demi and Howell, 1991). This potentially excludes them from opportunities for healthy adaptation to bereavement that can occur when families achieve direct communication and positive emotional interaction, when they become 'allies in grief' (Barlow and Coleman, 2003: 191). Findings from my own research indicate a lack of mutual expressions of grief within the family at the time of the suicide. Moreover, not sharing memories associated with the dead person is also a familiar theme in parental suicide bereavement.

Given the lack of social norms governing people's responses towards those grieving after a suicide, suicide-bereaved people may face stigmatisation, rejection and a lack of legitimation and sympathy that can inhibit their freedom to express grief (Corr, 2002). Within the bereaved child and their family's wider social networks, a disenfranchisement of grief is therefore likely to occur, potentially resulting in the withholding of social support by relatives, friends and neighbours (Doka, 2002). Whatever stresses pre-existed the suicide, little can prepare survivors for the long-term impact and complexity of their loss experiences (Wertheimer, 2001). Suicide bereavement thus poses quite specific difficulties arising from the suddenness and mode of death, the discovery of the body, police investigation, media reporting and the deceased parent's mental health (Hauser, 1987; Silverman *et al.*, 1994–1995).

How emotional and social turmoil of this kind can be made sense of theoretically is an issue addressed within both psychotherapeutic and sociological literatures which, respectively, have generated the concepts of the assumptive world and ontological security. Psychiatrist Parkes (1996) argues that one dimension of grief is the rupture of the bereaved person's 'assumptive world'. In other words, grief is not simply a matter of private pain. It resides within a bereaved person's previously taken-for-granted relationship with their external social and physical world. The concept of the assumptive world thus explains how people come to view everyday life as largely secure and predictable. As such, it helps to organise the way they relate to others, supporting their day-to-day encounters through beliefs about the benevolence and meaningfulness of the world and worthiness of people (Janoff-Bulman, 1989). Death, and in particular suicide, can shatter someone's assumptive world, disrupting a sense of autonomy and so producing anxiety (Janoff-Bulman, 1992; Valentine, 2008). The fact that a parent could kill themselves potentially erases children's trust in those they feel close to; their sense of self can fragment, leaving them feeling vulnerable and powerless (Janoff-Bulman, 1989; Kauffman, 2002).

Sociologists such as Giddens (1991) and Bauman (1992) have addressed this dimension of personal and social life via the concept of ontological security. This too encompasses trust in the orderliness and predictability of everyday life, particularly a sense of oneself as a coherent and durable being, despite transitions and crises. As Charles Taylor says, 'In order to have a sense of who we are, we have to have a notion of how we have become, and of where we are going' (1989, cited in Giddens, 1991: 54). Many suicide-bereaved people struggle to construct this sense of coherent continuity about '*who they were*' before the suicide, whilst also having to integrate into their daily existence, 'who they are now'. The 'who they are now' is someone marked not only by the legacy of suicide, but often by unhappy memories, mental ill health, fragmented family relationships and isolation. For some, their inability to trust other people and the world may cast a long shadow over their lives and personal relationships. It is therefore not just the inner emotional lives of suicide-bereaved people which are turned upside down; they are also thrown into an entire 'lifeworld' of chaos and turmoil (Fielden, 2003) which can leave a 'death imprint' potentially for life.

Faced with these circumstances, then, what resources might be available to suicide-bereaved people? As noted, emotional and social support within the family and its wider networks may be curtailed. However, the analytic concepts of assumptive worlds and ontological security not only illuminate apparently intractable circumstances, but also suggest an important role for new models of grief (Walter, 1996). As Walter notes, the (re)creation of biographical continuity can be an important vehicle through which bereaved people may, for example, construct the 'last chapter' of the life they shared with the deceased person. Whilst orthodox views of grief have been seen to privilege psychologised and medicalised models which promise 'healing' or 'cure' by severing bonds between the bereaved person and the deceased person, the notion of biographical repair contributes to newer models of grief that argue for the value of 'continuing bonds' between the living and the dead (Klass *et al.*, 1996; Stroebe and Schut, 2005). To afford the dead a continuing social presence in the lives of the living, then, bereaved people may 'make meaning' through conversations that involve biographical 'emplottment', that is, stories of the person who has died and stories of themselves (Árnason, 2000: 189).

That said, accounts of biographical repair or narrative transition (Neimeyer, 1999) often represent these as *social* practices; stories of difficult events are rehearsed with people undergoing the same or a similar transition. Neimeyer (1999), for example, describes the restorative qualities of talking to a 'fellow traveller'. As noted, however, communication between 'fellow travellers' may be precisely the resource that someone bereaved through suicide lacks. Yet what my data show are strategies of biographical repair which do not privilege the social medium of conversation. By privately retracing their parents' last hours Rose and Will were able to creatively 'story' their emotions and their relationship with their dead parent in an embodied, materially grounded sense. As their accounts suggest, this embodied activity proved cathartic, furnishing them with new ways of understanding and relating to the world. Thus, in the quest to create order out of painful and disordered memories, opportunities for meaning-making through 'pilgrimage' to the place of death or other emotionally significant locations potentiated a re-evaluation and transcendence of loss. Rather than a pilgrimage in any religious sense, such a journey holds symbolic and emotional significance for the traveller. However, while the notion of biographical repair (Walter, 1996) reflects a logo-centric society's verbal or textual models of meaning-making, an interpretation of Rose and Will's materially grounded accounts demands theoretical resources which can unpack long-standing relationships between bodies, both alive and dead, and the 'stuff' of landscapes and objects.

From within emotional geography, Davidson *et al.* (2005) have explored the ways in which landscape, places, objects and emotions both materialise and spatialise emotions. In the cases of Rose and Will a conscious re-evaluation of a landscape in which their parents had taken their own lives was achieved by quite literally 'walking the walk'. The space and landscape in which the suicide occurred was thereby transformed into a *place* invested with human meaning (Relph, 1976).

Methods

The study that informs this chapter used narrative research methods to allow participants to explore childhood bereavement experiences and family dynamics which might hitherto have been censored (Rowling, 1999; Valentine, 2008). In so doing, participants also recounted life-affirming experiences which had enabled personal growth. Of the 28 participants, 24 were parentally bereaved by suicide, or a self-inflicted death (coroners' verdicts included suicide, open, misadventure, accidental death or unknown). Of the other four, one was a lead consultant paediatric psychologist, the other two (also bereaved by suicide but not parentally) facilitated self-help suicide bereavement support groups. The final participant was my transcriber who had not been bereaved by suicide. I wanted to find out whether the sensitivity of the data had impacted on their understandings of suicide, and indeed whether the work had been emotionally costly.

Recruitment of a sample was challenging. Only four suicide-bereaved participants were male, reflecting the under-representation of men in bereavement research generally (Barrett and Scott, 1990; Range and Niss, 1990; Biddle, 2003; Mitchell *et al.*, 2003). The stigmatised nature of mental ill health, suicide and the subsequent trauma experienced by many survivors also inhibited potential participants (Table 12.1).

Participants' transcribed narratives were coded for themes and then cross-coded into categories. Crucially, participant validation has contributed to my analysis, and all members of the sample were offered the opportunity to feedback on publications, including this chapter. For a number, this was viewed as a positive aspect of the research process, as Rose noted:

> *... it's good to see everything in print, it makes my mum important ... .you understand what I mean by that.*

Table 12.1 Profile of the final sample

Child (under 18)/adult when lost parent	Length of time since death	Parent lost
12 participants lost a parent in childhood or as a young person	Ranged from 11 months to 53 years	10 mothers 14 fathers
12 had lost a parent as an adult, although it is important to note that five of those were bereaved between the ages of 18 and 21 and were still living at home, with having one participant only having recently left		

Case studies: Rose and Will

Rose was 54 and her mother's suicide had occurred 23 years prior to our interview. The youngest of five children, she had two elder brothers and sisters and described a working-class upbringing and a close family life that encompassed both happiness and difficulties. Despite pleasurable memories of going on walks with her mother, who liked to visit churchyards occasionally, there were periods when her mother 'disappeared' into a mental health hospital, a source of fear that permeated Rose's childhood:

> *So fairly often she wasn't there and either mentally, or sometimes physically and I used to like it when she was around. You know, I stuck like glue to her because I was – didn't want her to you know – sometimes she wasn't there and I loved her very much and she would take good care of me and she was lovely – and so, we were really close.*

A few days before Rose's 21st birthday, her mother removed herself from a mental health institution and after a couple of hours' walking, stepped into a busy road. Initially, the police constable did not tell Rose how her mother died, but she quickly became aware that her mother had killed herself:

> *...she'd walked 5 or 6 miles into the town centre and she laid down in the road and a truck ran over her.*

Rose has no memories of crying at that time, but 15 years later she experienced a 'grief spike' (Russac *et al.*, 2002) stimulated by her pregnancy and the birth of her only child:

> *Well, I kind of realised something wasn't right – well, when I got pregnant.... and that was when I realised that she wasn't there. And that's when I realised how much I wanted her... all the thoughts that I guess should have happened when she died and had been pushed away. They all – I couldn't stop any of it. It was all coming out.*

Rose recalled 'bad feelings' about her pregnancy and the child: '*I was so sure that it was all going to go wrong because anything I wanted was taken from me.*' Fear that her child would be hurt or taken away overtook her. Three years later, Rose briefly tried bereavement counselling but her feelings about her apparent inability to adapt to her loss was affecting her parenting skills and ultimately, the emotional well-being of her child. As a result she chose to do '*something about the situation*', undertaking what Cheal calls a 'ritual of resource management' (1988: 284) in which a ritual experienced in the present acknowledges the limitations of the past in order for the future to begin. Therapy was Rose's choice, a strategy which, she said, enabled her to 'walk the walk'. During the year prior to our interview she had:

> *...spent lots of time during my therapy thinking about 'the walk' that my mum did, the first time I had really faced it, I suppose, and I realised that I would have to walk it myself, so that I might be able to see it from her eyes, and somehow be in her shoes. To somehow get inside her head, to see the same things that she saw, and hear the same things that she heard – see the same trees, albeit grown, smell the same smells, and in a way I felt I was walking beside her all the way – I think I talked to her whilst I was doing it. My therapist didn't mention it, but I only had the guts to do it because I could walk straight in to therapy – literally – after the walk.*

This was not, therefore, an opportunistic or impulsive walk. Rose had planned it in great detail, down to the bus times.

Another participant who stepped outside of their everyday life to embark on an emotionally charged pilgrimage was Will. At 52, 40 years after his father's death, he retraced the final hours of his father's life. Will described his childhood as largely safe and secure; he lived a middle-class family life with his parents and younger sister and received a private education. Unlike Rose, who had had an acute awareness of her mother's mental ill health, Will did not know that his father was depressed. When he was 12, his father dropped him off in his car at school, went to work and then in his lunch hour drove to a nearby wood and hung himself from a tree. Will was initially told that his father had died in a road traffic accident, but 9 months later he discovered:

> *... there was a tin box under the bed and I got the tin box out and papers of my mother's in there and I found my father's death certificate. And it said he took his own life, so I knew.*

He went on:

> *So I was conscious, even before I knew he had taken his life and that it had been in the papers. Because, one of them came up to me, one of my school friends or school acquaintances saying, 'it was your father wasn't it, that died it was in the papers?' and I said, 'yes'. But so, clearly he knew more than I did. I didn't know anything about how he died – nothing.*

As already argued, bereavement through parental suicide can radically overturn individuals' assumptive worlds, leaving them with a lasting sense of ontological insecurity. The case studies presented in this chapter thus describe two individuals' experiences of a parent's suicide and their much later responses to those circumstances, for example, in Rose's case as a result of difficulties with her own parenting.

Finding order within chaos

To what extent, then, can Rose and Will's actions be understood as attempts to repair the rupture of an assumptive world, to restore a sense of ontological security? Biographical repair, as a dimension of new models of grief (Walter, 1996), has been posited as potential resource for individuals bereaved in this way. Yet the emphasis on conversation, the need for 'fellow travellers' (Neimeyer, 1999) militates against the deployment of verbal strategies. Instead, it would seem, embodied engagement with the emotional geographies of suicide and loss offer another way forward. Working with the data presented here, however, it is useful to consider both the extent to which these participants' choices constitute an example of this strategy and the processes through which emotional and social transitions might have been achieved.

My use of narrative interviewing allowed participants to not only recount but also reflect upon their histories. Data therefore encompassed the ways in which they themselves made sense of both their distant and recent pasts and these interpretations provide an important analytic starting point. As Lakoff and Johnson (1980) argue, human beings utilise apparent similarities between different domains of life in order to extend their understanding of chaotic, unfamiliar or more abstract experiences. In other words, familiar experiences – such as going for a walk or making a journey – can help frame therapeutic 'excursions' into unknown emotional 'terrain'. In the case studies presented here of Rose and Will, material metaphors such as walking and journeying were not only deployed during an interview, but also given tangible form in embodied, spatialised practices. In this respect, what Rose and Will undertook was a ritualised pilgrimage, in that ritual is a site at which metaphoric thought is given substance, enacted in a particular, often sacred, time and space. It is in this way that it becomes persuasive, so enabling transitions to occur (see Hockey, 1990: 84–85).

Viewed from this theoretical perspective, the spatial metaphors of 'walking the walk', 'sentimental journey' and 'the re-enactment' not only structured Rose and Will's thinking but gained persuasive power through their ritual enactment at times and spaces for long made sacred, or marked off from their everyday lives. As such, these materially grounded, spatialised metaphors had the power to construct reality (Lakoff and Johnson, 1980); as embodied narrative devices, Rose and Will's metaphors became tangible vehicles that allowed them to make sense of their loss and (re)generate their memories of their parents. Importantly, the ways in which these metaphors were deployed demonstrates the symbolic separation of particular experiences of walking, visiting a wood or taking photographs, from the otherwise mundane nature of such activities.

As a conscious self-reflexive reordering activity by both Rose and Will, the materiality of objects, environment, texts and conversation were being used to form a connection and bond. As a set of practices, they potentially preserved and (re)constructed a relationship with the deceased parent in the face of bodily absence (Hallam and Hockey, 2001; Hockey *et al.*, 2005). In negotiating and renegotiating understandings of his father's death, however, Will found that his initial decision to undertake a 'sentimental journey' led him into unplanned, spatialised activity:

> *I'll try climbing this tree and I caught myself thinking, 'This is crazy. What are you doing? You know, what are you doing in the tree?'*

Via the wood (Figure 12.1), the tree, his emotions and his body Will was able to engage with the 'emotional spatialities' of his bereavement

Figure 12.1 A site of death: The wooded area where Will's father was found hung.
Source: Will, 2002/03.

(Jones, 2005: 205). He went so far as to climb into the tree he guessed to be the one where he imagined his father had been found hanging in his work suit and overcoat; this image of his father's body clothed in this way had held a grip on Will for over 40 years. Neimeyer (2000: 550) describes such practices as potentially raising 'new questions' and undermining 'old answers'. For Will, his re-enactment indeed undermined old answers and, while raising no new questions, provided new answers to some old ones:

> *...when families aren't honest with each other. I mean, my mother put it, presumably to protect me, that you know, it was as though something had just overtaken him. Some madness has over-taken him.*
>
> *The other thing I discovered, because I even went into the wood and looked for the tree and the other thing I discovered was that, he – I – looked at it and I thought, 'He's not done this suddenly.'...so by doing the re-enactment I discovered that it was planned.*

That his father had left work and taken a rope to the wood and been able to climb a tree wearing a suit and overcoat demonstrated to Will that the suicide had been planned – his father had been determined to end his life. This discovery alerted him to a biographical incongruence between what his mother had told him and what Will believed had happened. Thus the re-enactment facilitated Will's biographical continuity:

> *I have this information. He made the decision to walk away from us in cold blood.... I was able to leave my father in the wood and complete this aspect of my business with him – even though I hadn't found out everything that I could have found out. Afterwards, when I wrote about this experience, I put it like this, 'enough is enough, I'll leave my father to haunt the place.'*

His choice of language reminds us that deceased parents can act as 'social ghosts', exercising agency even in death, influencing, guiding and encouraging their children (Gergen, 1987 cited in Stroebe *et al.*, 1996; Valentine, 2008). However, Will describes a diminished sense of his father's power as a social ghost after the re-enactment. Although the discovery of a profound misalignment between his memories of his father's suicide and the apparently rational decision his father had made to kill himself was unwelcome, it nonetheless served to redefine the way Will interacted with the memory of his parent.

As argued, therefore, the embodied enactment of metaphors within ritual times and spaces, set apart from both everyday life and, in these cases, the preceding life course, acts to affirm the persuasive power of a metaphoric framework. Throughout this chapter, however, it has been argued that bereavement through parental suicide has a set of quite distinctive characteristics which can *disempower* children, who already stand in a subordinate, dependent relationship with their parents. Thus stigma, coupled with a conspiracy of silence and a censoring of grief, presents barriers to bereavement adaptation. In addition, however, a parent's suicide is a final act which is 'acted upon', rather than children, whether young or old, 'acting in' their parent's decision (Richert, 1999). Here the concept of 'acting in' is used to suggest a level of agency that involves consultation, collaboration and agreement between parents and children. Yet bereaved children are 'acted upon' since their parent's autonomous decision to kill themselves immediately severs the parent and child relationship, and may also throw bereaved children into a lifeworld of chaos and emotional suffering. In that their social location means that children lack agency when compared with their parents, for Rose and Will retrospectively 'acting in' their parents' last few hours was an empowering experience. Rose said what she wanted to do was:

...walk the walk with her, not have her be alone when she did it...as if I was pretending that this was the first time, the real time, when she really did it, I didn't want her to be alone when she died. That was what I really wanted to do – be there for her (which I wasn't, of course). I wanted to experience what she did.

(Rose)

In this way Rose attempted to 'act in' her mother's suicide. Will too 'acted in' his father's last few hours, something he conceptualised as 'the re-enactment'. He later described this in his journal entitled, '*Sentimental Journey*', noting that the title was an ironic literary allusion because at that time he felt anger towards his father. Like Rose, he planned his re-enactment in some detail as he had to travel to where he had lived as a young boy in the 1960s:

So I went to my old house where we lived and then I went to the school, which is a significant memory for me because I can remember the car driving off and I went to the school and he went to work and then I went to the place he used to work.

As active agents both Rose and Will were able to address both the past and present in order to accumulate memories and evoke feelings. This they were able to achieve by developing new orientations through a pilgrimage towards the suicide and the events surrounding it. As pilgrimages of restoration, 'walking the walk' and 'the re-enactment' enabled them to physically enter the emotional landscapes of their parents' deaths and experience their own embodied, emotional responses to these landscapes. Hitherto, Rose and Will had not visited their parent's site of death as these spaces had been invested with negative associations. It was during their pilgrimages that the spaces where their parents had walked, worked and subsequently died became transformed into *places*. As discussed earlier, place, as it was experienced by both Rose and Will in a reflexive and self-conscious way, is distinct from space (Relph, 1976). Thus, whilst their parents died in public spaces, to Will and Rose, these spaces were private emotional places invested with personal meanings of pain, suffering and loss.

Repairing the assumptive world/restoring ontological security

Thus far this chapter has shown how the enactment of embodied metaphors which drew participants back into long distant emotional geographies can work to reshape their conceptions of what had occurred, so enabling emotional transitions and restoring ontological security. With regard to the new 'assumptive worlds' into which both Rose and Will were moving, however, insight might also be gleaned from considering the memorialising dimensions of their practices more directly. In other words, the (re)creation of memories of their parent was a future-oriented practice. As Jones (2005) argues, in spatialising emotions, memory plays a 'formative role in the construction of our *ongoing* emotional and imaginative geographies' (p. 210, emphasis added). Memory is the bridge between a world that *was* with the deceased person and is *now* without them (Buchsbaum, 1996). Images, individual feelings, the death itself and other peoples' responses come together to create a memory of the time around the suicide. Yet as argued for many suicide-bereaved people, especially those bereaved as children, the retrieval of memories can be very difficult. Findings from the study presented here indicate that the 'blocking out' of events is a frequently used self-protective mechanism (Parkin, 1993). This failure to remember, or 'blocking out',[2] meant that a number of participants had repressed their memories in order to cope with distressing experiences:

> *And I think, to me, part of – a lot of stuff either went over my head, or I ignored, or I just put it away somewhere that is not retrievable – 100% blocking off all of it. Every feeling. Every memory. Every arrangement. It's gone. It's not there for me.*
>
> (Rose)

Years later, as a strategy for retrieving and creating memories as well as integrating her mother into her life, Rose commented on how she:

> *. . . wanted to do the walk she [her mother] had done . . . . I remember wondering if her hands had been crushed, hands that had nurtured and held us all, that had been tender and strong. It made me wonder how many bones had been broken, and how many people had seen her in the road. I remember hoping that her dress had not ridden up, that she was 'decent', as she would have put it because she was a very proud and private person, and would have been mortified to think her underwear had been on show.*

'*Walking the walk*' was not only a reminder of her mother, but also allowed Rose to construct her own picture of what may have happened (see Walter, 1999). Both Rose's 'walk' and Will's 're-enactment' actively evoked remembered and forgotten memories and emotions which then brought their parent's absent presence into focus (see Hallam and Hockey, 2001):

I wanted to know what was going through his mind on the day he died and I wanted to also evoke how I was feeling on the day he died, because he just disappeared.

(Will)

Will took further steps to create a 'durable biography' (Walter, 1996) of his father and himself. In enacting this 'last chapter', he took photos of his pilgrimage – of his old family home, school and the wood in which his father died. As noted by Riches and Dawson (1998), when viewed photographs can become a focus for emotional release and a vehicle for negotiating a social bond, they also connect us materially, emotionally and memorially with the deceased person (Gibson, 2008). Yet with the passing of time, memory may become neutral and fade – it can shift from colour to black and white, and then to sepia. Will's photographs provided concrete evidence of his father's last few hours and enabled him to visually recall his father and the suicide. As already noted, Will's father driving off in his car is a significant memory as it was the last time he saw his father alive. During his 're-enactment', Will took a photograph of a car driving off outside his old school as this was his last memory of his father (Figure 12.2).

Since photographs lack immediacy and look 'backwards in time' (Gibson, 2004: 290) by taking these photographs, Will can be seen to rewind time back to his father's death and blur the distinctions between the living and the dead. The result is that he was able to 'be there' in his old life and also 'be in' his father's last few hours. On the one hand, these photographs documented Will's trip, but on the other hand, the photographs provided a metonymic, or part for whole, link with his father, potentiating a sense of his presence (Riches and Dawson, 1998) as a visual representation of Will being 'there' and 'in' his father's last few hours.

'Being there' and 'being in' were also evidenced in Rose's walk in that her feelings and conversation with her mother at the place of death became a site for Rose to sense her mother's 'presence and agency' (Valentine, 2008: 151).

And, I could see her standing by the road. She was right in front of me.... and I couldn't run – I didn't run, but walked really quickly towards her and then she was gone. She wasn't there. She was gone.

In this instance, the vision of Rose's mother at the site of her death constituted a positive and welcome presence, and one that Rose took comfort from. However, medical discourse might view her vision or

Figure 12.2 Connecting to the final memory of his father: A car driving by Will's old school.
Source: Will, 2002/03.

'hallucination' as pathological, a symptom of acute grief or mental ill health (Becker and Knudson, 2003). It is also possible that this extra-sensory perception of Rose's mother may have resulted from chemical changes in the brain due to her heightened emotions at the time of the walk (see Bennett and Bennett, 2000). Nonetheless, whilst the dead are invisible, they do exist as images in our minds. These images can appear outside our minds and subjectivities as 'animating powers of the imaginal world' (Becker and Knudson, 2003: 714). Indeed, Marwit and Klass (1996: 298) propose that the 'inner representation of the deceased' plays an important function in the ongoing lives of bereaved people. Evoking the parent's presence is an integral component of maintaining a continued relationship and bond (Marwit and Klass, 1996). For example, whilst 'walking the walk', Rose talked to her mother – and indeed talking to the dead is not an uncommon practice (Klass *et al.*, 1996; Bennett and Bennett, 2000), whether the bereaved person is awake, semi-conscious or in dreams (Bennett and Bennett, 2000; Howarth, 2000).

In blurring the boundaries between the living and the dead, by keeping memories alive, Rose's interaction with the inner representation of her mother's presence facilitated what she perceived to be an active and healthy response to her loss. Perhaps as a 'last chapter', *'walking the walk'* constituted a spatialised focal point in which the living (Rose) and the dead (her mother) were briefly and intensely connected. Even in death, then, as a social ghost, Rose's mother continued to influence her thoughts and actions in a positive sense. Rose's account of 'walking the walk' can therefore be seen as a narrative co-constructed and indeed, co-authored, with her mother. In mediating a continuing bond, both Rose and her mother were writing and enacting the walk and indeed, they then storied this to an 'outsider witness' which was me, the researcher (Payne, 2006: 16).

At the same time, her mother's presence also evoked her absence and whilst Rose remembered memories and connections with her mother, feelings of loss were also generated (see Hallam and Hockey, 2001). As Rose has noted above and elsewhere during the research interview, she felt sad and tearful (and did cry during the interview) when recounting her bereavement experiences. At the same time, there was an acceptance of her loss and an understanding of her mother's last fatal action. Ultimately, walking in her mother's footsteps enabled Rose to make sense out of a seemingly senseless and meaningless act:

> *But to me, doing that walk was basically a good thing to do and it got a lot of things straight in my mind...*

Conclusion

While western societal attitudes towards suicide are shifting, it is still potentially perceived as a stigmatised death (see Chapter 6), a status further compounded when suicide bereavement is not publically acknowledged or socially validated (Corr, 2002; Armour, 2003). As noted, suicide can discourage familial and social communication with the result that grief can become hidden, thus inhibiting bereavement adaptation. Given that the suicide may have been the culmination of years of difficulty, often marked by previous suicide attempts, threats of suicide and familial disruption, family communication patterns may already have been disrupted before the death. Communication can be further undermined by feelings of rejection, blame, shame, anger and real or imagined stigma, all of which are barriers to positive and open family communication and interaction.

As a result, suicide can leave bereaved people feeling vulnerable and fragile. For the parentally bereaved, particularly if the suicide occurred during their childhood, psychological pain and suffering may cast a long a shadow over their subsequent lives. The aftermath of suicide is often overwhelming and whilst sense-making *may* occur shortly after the death or across subsequent years, it can also be decades later when bereaved people are able to turn and face the shadow of suicide and its subsequent losses. The research project drawn on here contributes to an understanding of these circumstances, ones which have been neglected in academic and therapeutic literatures. It provides a framework for making sense of how loss, emotions and landscape may be used to find some order within the disorder generated by a suicide. Rose's 'walk' and Will's 're-enactment' were undertaken in emotionally significant places that were marked off from the practical aspects of their everyday life and accorded a special status by them (Vergunst, 2008). As active agents, Rose and Will were able to develop alternative understandings though the spatialities of their emotions in what they experienced as a restorative pilgrimage. That people consciously use the landscape to create meaning and a connection with dead people is not a new phenomenon, and interaction between the living and dead, previously framed as pathological or deviant, may alternatively be viewed as adaptive and therapeutic.

Whilst the landscape can never be a substitute for the dead person, this chapter has illustrated how suicide-bereaved people may nonetheless use a space which is invested with pain, suffering and death to interact with a deceased parent. This space can then be transformed into a place endowed with human meaning, and one which allows the dead person to have a continuing social presence in the lives of the living. The public spaces where their parents died were thus experienced by Rose and Will as private symbolic places which were profoundly and uniquely meaningful to them.

Notes

1. Rose and Will are pseudonyms.
2. Several participants conceptualised their lack of memory as 'blocked out'.

13

Potent Reminders: An Examination of Responses to Roadside Memorials in Ireland

Una MacConville and Regina McQuillan

This chapter addresses the practice of establishing memorials at roadsides, demonstrating their potency or symbolic efficacy, as items of material culture, to mobilise powerful emotions such as grief, fear and anger. Through a range of examples, it explores the contingent nature of these emotions, deriving from temporally located personal and political relationships with the deceased person.

The placement of flowers, cards and mementos by the side of the road has become a recognisable sign of a sudden or tragic death. Fatal accidents, large-scale disasters or tragedies frequently elicit this kind of response, and Santino (2001) has coined the phrase 'spontaneous shrines' to mark this form of material response to an untimely death. According to Santino, these shrines broaden the response beyond family and friends to a larger public audience and as such they are communal responses to tragedy where anyone may add a token if they wish. These 'shrines' are typically considered not only as an appropriate but also as an expected response to tragedy and express attempts to make a catastrophe, on a large or small scale, more manageable. The creation of a shrine can give people a sense of purpose (Grider, 2001). They 'emerge quickly, and often within a few hours of an event' and stay in place throughout the liminal period between death and burial; after the funeral, when shrines are removed the sacred space returns to a secular space and the graveyard becomes the place to leave mementoes (Grider, 2001: 2). Often the only traces that remain are decaying flowers and rain-sodden cards.

However, in some cases more permanent memorials are erected at sites of sudden or tragic death and it is these that have now become a focus of study, documented throughout Europe, Canada, Australia, and North and South America. In some countries they seem to be a relatively new phenomenon, particularly so in Australia where they do not appear to have been present prior to the 1980s (Clark and Frauzmann, 2002). Elsewhere they seem to be part of a continuing and probably Roman Catholic tradition of remembering the dead and marking death in open places (Grider, 2006). They provide potent reminders of tragic intimate experiences, albeit in public places (Collins and Rhine, 2003), and generally mark road accidents, although a considerable number also mark suicides, deaths by drowning and other tragic deaths.

Responses to these memorials, both private and public are varied, and are rarely neutral. Frequently they can meet with strong opposition; for example, they are considered illegal in several States in the United States. Research in Canada (Churchill and Tay, 2008) suggests varied official and unofficial responses to these memorials with State and local authorities demonstrating awareness of their sensitive nature while trying to balance issues of driver and road safety. Attempts to get the balance right regarding material expressions of private grief within public space has prompted Collins and Rhine to query whether in the matter of death, which is the greater right – 'to grieve and mourn or to be spared reminders of our own mortality?' (Churchill and Tay, 2008: 241) (see Chapter 8).

In the case of Ireland, a more diverse and complex set of memorialising practices are evident at roadsides. In the Republic of Ireland, they can be considered a continuation of a *tradition* of marking death in open places, albeit often in a modified material form when compared with the past. They are numerous and are generally accepted by neighbouring communities – in many cases local authorities even facilitate the erection of memorials. This is in contrast to the reaction to some memorials in parts of Northern Ireland, a political and physical topography containing differing, and frequently conflicting cultural traditions. The roadside memorials erected to people who have died as part of the political turmoil, on both sides, are of a different nature to those erected elsewhere to victims of road accidents. However, the vociferous responses to these memorials in Northern Ireland demonstrates the potency that resides in this material form and raises questions beyond issues of grief and personal mortality.

In this chapter, we will consider these memorials within a wider context – their forms, functions and the responses to them – in order to more fully understand the complex responses to memorials within the island of Ireland.

Contemporary roadside memorials

Before considering the disruptive nature of these memorials in Northern Ireland, understanding the purpose of, and responses to, roadside memorials in places other than Ireland demonstrates that even in non-political contexts these memorials can be troublesome. A number of contemporary studies which have interviewed family members and friends of the deceased person (Hartig and Dunn, 1998; Santino, 2001; Everett, 2002; Excell, 2004) suggest that these roadside memorials have two main purposes: warning and remembrance.

Remembrance and continuing connection

Roadside memorials can be deeply personal sites of remembrance that provide continuing connection with the deceased person. For some family members and friends the place of death and the site of the memorial become the location for a continuing dialogue with the dead (Everett, 2002). One family member speaking on an Australian radio programme explained that the roadside memorial marked '*where he lost his life and where he will always be ... I won't go out to the cemetery because that's not where he is ... he is here at the cross*' (ABC Australia, 7 December 2003). Although Grider (2001) has argued that 'spontaneous shrines' become a temporary 'sacred space' and return to secular space once burial has taken place, for many of these family members, it would appear that the place of death *remains* a significant space where the living may feel closest to the dead. As sacred spaces, the memorials may thus become a focus for pilgrimage, for months or years, being not only visited but also maintained. Clarke and Cheshire (2003: 214) state that there is 'ample evidence to indicate that many roadside memorials are revisited on holidays, the birthday of the deceased and the anniversary of the fatal crash' – for example, at Christmas memorials may be draped with tinsel or hung with seasonal decorations.

These practices can be seen as an ongoing dialogue, as one memorial poignantly illustrates. From a study of roadside memorials on one major route way, N4, in the country we recorded a delicate metal cross, erected to a child who died in 1949 aged 9 years. While it is not possible to say when this was erected, its style and form would suggest it is not a recent artefact. Fresh flowers were placed at the memorial on the occasion of

Figure 13.1 Roadside memorial erected to a child who died in 1949; flowers on the 55th anniversary were recorded.

her 55th anniversary, clearly demonstrating that this is an enduring site of remembrance and continuing connection (Figure 13.1).

In a small American study of 14 interviews with people who have erected roadside memorials, Collins and Rhine (2003: 228) found that nearly all of those interviewed rejected the notion that the purpose of these memorials is to 'say goodbye'. Instead, for them, the emphasis was on continuing the connection, with frequent references to 'not letting go'. Collins and Rhine (2003: 214) argue that these memorials signify abrupt and unexpected deaths, that those left behind are flooded with emotions – guilt, loss, impotency and separation – which drive attempts to mark the spot where people have died and illustrate a sense that the deceased person 'sees' or 'knows' what is taking place at the site of death.

Memorials as warnings

Roadside memorials and 'spontaneous shrines' also operate as an effective warning for road users of the physical dangers of the road. The front cover of the *Sunday Times* magazine in 2004 featured a roadside 'spontaneous shrine' to illustrate a feature article on the dangers of the A59 road in England (26 September 2004). A UK charity, Roadpeace, dedicated to supporting bereaved and injured crash victims and founded in February 1992 (www.roadpeace.org), developed a programme of providing standardised memorials to be placed at the roadside by friends or family bereaved through a road traffic accident. This programme – 'Remember Me' – has erected over 2000 markers since its launch on 31 August 2003. These markers are black A4-size signs depicting a red anemone and are intended to be noticeable, non-denominational reminders of the dangers on the roads. In the United States, Mothers Against Drunk Driving (MADD) have erected roadside memorials to those who have died, especially as a result of drink driving. The families involved saw part of the purpose as educating the public about the dangers of this road with memorials serving as a warning to other road users (Everett, 2002: 104–7; Arnason, 2009).

While these might appear to be recent initiatives, Clarke and Cheshire (2003) record evidence of crosses being erected by road authorities to mark particularly dangerous intersections in the United States in the 1950s and 1960s. Although these were not personal spontaneous memorials, their purpose was to advise travellers to drive carefully because a large number of people had died at this point in the road.

Opposition to roadside memorials

As multi-vocal symbols of personal loss, set up in public spaces, roadside memorials are not therefore neutral artefacts; while they can be of huge significance for relatives and friends, they are not universally accepted. Concerns are expressed that memorials serve as a distraction on the road thus becoming a further danger. In the United States, attitudes differ across the country and in some states they are banned. For example, Wyoming, which has one of the highest traffic fatality rates in the United States (Frosch, 2008), in 2003, enforced a ban, after their proliferation was considered to be a distraction. However, currently, the state of

Wyoming now produces and maintains its own uniform signs, depicting a dove over a broken heart and these are left in place for 5 years. In other states, such as Alaska and West Virginia, memorials are facilitated if they follow particular procedures, including guidelines about size and position (http://www.transportation.wv.gov/highways/traffic/roadsidememorials/Pages/default.aspx and http://www.dot.state.ak.us/stwddes/dcsrow/assets/pdf/roadsidememorials.pdf).

Recent surveys in Canada (Churchill and Tay, 2008; Tay, 2009) found that among 82 municipalities surveyed a vast majority (90 per cent) did not have a policy in relation to roadside memorials but generally preferred to allow them, albeit with some restrictions. The concerns of the municipalities focused on issue of driver distraction, safety, maintenance and liability while being aware that these memorials may help people grieve for someone who had died.

In the United Kingdom, there are similar concerns about safety issues surrounding these memorials. At least four local authorities have told bereaved families not to lay flowers at the site of a fatal crash after a predetermined period of mourning; sometimes as brief as 2 weeks, and they are seeking a ban on permanent roadside memorials (Gadher, 2004). These restrictions are for 'health and safety reasons', and authorities are claiming that they are acting on advice from the police who fear the tributes, many near dangerous roads, could distract motorists. These restrictions have been considered 'crass and insensitive' by bereaved families and road safety campaigners, some of whom feel these memorials are being removed in order not to draw attention to the death toll on certain roads (Gadher, 2004).

In addition to concerns over safety, it is clear that roadside memorials disrupt and disturb notions of private and public, secular and sacred through their material form and location. Many roadside memorials take the form of the traditional Christian symbol of the cross. In the United States, organisations, such as American Atheists and Americans United for the Separation of Church and State, raised objections to the use of religious symbols on the memorials (www.freepublic.com), and the original cross memorials erected by MADD have now been modified. Robert Tiernan, a lawyer with the Freedom From Religion Foundation in Madison, Wisconsin, is reported as saying 'for us, the memorials raise serious church-state constitutional concerns because they usually feature religious symbols and are placed on state property... the public space belongs to everyone, and I think it's important to honor that' (Urbina, 2006). In 2001, a Colorado court took the view that the removal of roadside crosses by a member of the public was not illegal as the roadside memorials, being unauthorised, were 'litter' and not 'venerated objects' (Barker, 2001).

Roadside memorials in Ireland

The debates surrounding these memorials within highly individual and multi-cultural societies such as the United Kingdom and the United States focus on the appropriateness of forms of material culture located within the public sphere. In more homogenous and arguably less individualistic societies, such as the Republic of Ireland, there is an acceptance, and indeed facilitation, of roadside memorials. As noted, this may reflect a predominately Roman Catholic culture that has strong traditions of funeral attendance and remembrance of the dead. Indeed the acceptance of these memorials in other Roman Catholic societies has been documented. Grider, commenting on the descanos (roadside memorials) in New Mexico, argues that these are essentially manifestations of 'folk Catholicism' (cited in Alessi, 2006: 22).

Furthermore, it has been suggested that there is a preoccupation with death in Ireland, Witoszek and Sheeran (1994) likening its funerary rituals and practices to a '*Theatrum Mortis*' – a theatre of the dead. The various traditions, performed by the living (the actors) for the dead (the audience), reveal an underlying cosmology, which Witoszek and Sheeran argue, is 'an acute awareness of a powerful realm bordering on the world of the living' (1994: 7). These authors also consider that this view of the dead, one that sees the dead as occupying an almost tangible existence, is so embedded in cultural practice in Ireland that it goes unnoticed.

Roadside memorials can be considered as one aspect of this almost tangible existence of the dead, something that has been the case for a considerable length of time. For example, the practice of building stone heaps or cairns, which has also been documented in Scotland, commonly occurred in Ireland (Nic Neill, 1948; Gosling, 1993; MacConville and McQuillan, 2005). The marking of sudden deaths by the erection of roadside cairns, to which passers-by would add a stone and pray for the deceased person, was a practice within living memory when the Irish Folklore Commission conducted a survey on 'death markers' in the 1930s. According to some respondents to this questionnaire, these cairns were more commonly used to mark sudden deaths, and it was customary for passers-by to add a stone to the heap and to bless oneself and say a prayer at the site. Adding a stone to the heap was regarded as being of benefit to the dead person and was accompanied by a prayer. This interpretation was obviously influenced by Catholic beliefs in Purgatory and in prayers for the dead (Nic Neill, 1948). While the Church does not *require* these practices, they appear to be similar to the vernacular practices documented by Grider (2006) in New Mexico.

In contemporary Ireland memorials are common on all roads. A website (www.iol.ie/-roadside memorials) established with the intention of 'Helping make our loved ones universally known, mourned and prayed for' has documented almost 700 roadside memorials and is by no means an exhaustive list. Regularly updated, it references memorials dating from 1894 through to the present day.

Memorials in Northern Ireland

Roadside memorials have also been erected in Ireland to mark sites associated with political deaths, including wars, uprisings and executions. The number of Irish political roadside memorials increased in the twentieth century with The War of Independence (1919–1921) and the Civil War (1921–1923). A common feature in the landscape of the Republic of Ireland, for the most part, they commemorate distant events. In addition, they are located in a physical landscape whose ownership is no longer contested and marking death in open places is a culturally accepted practice. In Northern Ireland, however, the memorials erected to people who have been killed in recent political turmoil can evoke conflicting and highly charged responses.

Northern Ireland became a separate entity as a result of the Government of Ireland Act 1920. Its origins as a distinct cultural and political entity are complex, dating back to at least the late twelfth century; they derive from successive invasions and the British conquest of Ireland. The most intensive effort to change Ireland was in the seventeenth century, starting with the 'plantation of the six counties' in the north of Ireland in Ulster by Protestants from England and Scotland, with the intention of bringing Ulster under the crown. Land was confiscated from the Irish population and given to Protestants on condition that they bring in Protestants tenants and build defences.

Ireland continued to have rebellions against English rule from the seventeenth up until the nineteenth century. The War of Independence from 1919 to 1920 culminated in the Treaty and the Government of Ireland Act creating the Irish Free State and Northern Ireland, which remained as part of the United Kingdom, and consists of the six counties of the Ulster Plantation of James 1 – the majority population being Protestant and of English/Scottish descent.

In contemporary Northern Ireland, the legacy of the past resonates in the erection of and responses to roadside memorials. In their material form, these memorials are difficult to categorise. Are they roadside memorials within a recognised cultural tradition or war memorials? Placing memorials at the site where people died disrupts the conventional practice of memorialising the war dead in locations sanctioned by the War Graves Commission.

While war memorials in locations of this kind can cause disruption, there are exceptions. Those located in the communities of the people they commemorate tend not to attract the same level of attack. In fact, they can help sustain the identity of a particular community (Figure 13.2).

However, many such memorials are subjected to continuous attacks and acts of vandalism and this is especially so if they commemorate the deaths of people who are not from that community. The powerful emotions that are evoked by these memorials reflect the continuing struggle to find some level of reconciliation among Northern Ireland's different communities.

A motion was brought before the Northern Ireland Assembly in 2002 stating 'That this assembly rejects the offensive trend of erecting

Figure 13.2 Memorial commemorating five people killed in the Shankill Bayardo Bar, Shankill Road, Belfast in the heart of Protestant Belfast. Photograph David McCormick.

memorials throughout Northern Ireland by Republican elements in memory of terrorists who tortured citizens of this state for decades by their campaign of murder, maiming and destruction and calls upon the Executive to take immediate action to remove those memorials which have been erected without permission' (www.niassembly.gov.uk/record/reports). This motion sparked a heated debate. The proposer, Mr Foster, claimed that Republicans were seeking political gain at the expense of the relatives of the deceased, and strenuously argued that:

> *The motion is important to the many relatives of those who were foully murdered by the terrorists who held this land to ransom for many years. The erection of forms of memorial to those who for decades wrought havoc and destruction on the people of this state is, at least, highly offensive, in-your-face and profoundly insulting. It is uncaring, uncompassionate and grossly irresponsible. It is an offensive taunt to a community which has suffered broken hearts, broken limbs and broken homes and been left with heart-broken widows, mothers, fathers, brothers and sisters and many orphans. Sinn Féin still lauds, and associates with, the gunman and the murderer, despite the fact that its members now act in Government here in Stormont, as Ministers acting on behalf of Her Majesty The Queen in this part of her realm. (www.niassembly.gov.uk/record/reports)*

Another supporter of the motion, Mr Gibson, added that:

> *I do not call these objects 'memorials'. They are wayside platforms for the promotion of the political activists who have supported terror in this country. This is another part of the IRA/Sinn Féin terror campaign. It is a form of political and institutional terror and a way of dominating the landscape. (www.niassembly.gov.uk/record/reports)*

However, Mr Gibson did not criticise memorials to Loyalist terrorists. An amendment to this motion, tabled in this debate by the Social Democratic Labour Party (SDLP), 'to provide guidelines so that memorials of whatever kind conform to agreed criteria and do not give offence' was rejected by the Assembly and the original motion was carried.

Despite this motion, memorials continue to be erected and continue to be contentious. Susan McKay, a journalist writing in *The Irish Times* newspaper in 2006 lists some of the recent acts of vandalism on these memorials; for example, in Antrim the destruction of a roadside shrine to Michael McIlveen, a 15-year-old who was killed in 2006 – the written messages from his teenage friends at the memorials was destroyed by paint-bombers; in 1992 a memorial to eight workmen killed in Teebane by the IRA was vandalised many times and in 1996 it was rebuilt. As McKay states, 'this was an intimate war – neighbours set up and murdered neighbours, leaving bitterly divided villages and towns. Memorials are a source of pride to some in the community; pain to others'. Other examples included a man who believes his son was killed by the and Ulster Defence Regiment (UDR, part-time British Army Soldiers) has appealed to Lisburn town council not to erect a memorial to UDR soldiers who have been killed.

Responses to memorials in Northern Ireland highlight difficult issues but also reveal the potency of these artefacts. This raises several questions. How are members of the IRA or loyalist paramilitaries to be considered – as volunteers on active service or as terrorists? Are town centre cenotaphs that commemorate the deaths of British soldiers who fought in the Great War offensive to the nationalist population where British soldiers had killed civilians? Are the more elaborate graveyard rituals and roadside memorials of the Irish Nationalist Catholic communities, North and South, a legitimate cultural and religious expression? Are the legislative requirements to promote equality and good relations in Northern Ireland compromised by allowing the erection of memorials, which might be perceived to mark out territory and inhibit the use of space by all the community? These are not easy questions to answer.

Nevertheless, efforts have been made to address these questions. In the 1990s, the Northern Ireland Community Relations Council and the Arts Council of Northern Ireland commissioned a report in order to consider suitable forms of commemorating the casualties, issues and lessons of conflict. Jane Leonard, its author, considered memorials in other countries and particularly those relating to civil war. In August 1994, a national memorial was proposed within a few days of the IRA cessation of violence. Early proposals included a wall on which would be recorded the names of victims in date order similar to the Vietnam Veterans Memorial in Washington, DC.

The intention of the Vietnam Veterans Memorial was, according to the architect Maya Lin, 'to bring out in people the realisation of loss and a cathartic healing process' (Leonard, 1997). While American public opinion was deeply divided over the ethics of participation in Vietnam, the memorial nonetheless had the straightforward task of commemorating one side's casualties – those who died in American uniform.

The problem remains of how to remember the dead from different communities in Northern Ireland. If a peace sculpture is ever to be commissioned, Leonard argues, its role, in expressing the grief and reconciling the post-war hopes of different bereaved communities within and beyond Northern Ireland, will be far more complex and challenging than the Vietnam project.

The following expressions of loss, desire for revenge, national and religious identity, bewilderment and continuing vulnerability on some existing memorials provide stark evidence of the complexity of dealing with these memorials and the ongoing disruption they cause.

> *In memory of Robert James Harrison, a faithful and devoted member of this lodge for many years who was brutally murdered by the enemies of Ulster on 5th February 1971 whilst serving his Queen and country. Gone but not forgotten.*
>
> *To the memory of Private William Robert Davies, Royal Regiment of Wales. A son of this village. Killed at Lichfield, June 1st 1990. During his 18 years he contributed richly to the life of this community in sport and companionship.*
>
> *In proud and loving memory of one of Ireland's noblest sons, Captain Michael McVerry, O/C 1st battalion, South Armagh Brigade, Óglaigh na hEireann, born at Cullyhanna 1 December 1949; killed in action by British forces of occupation at Keady RUC barracks, 15 November 1975.*
>
> *In remembrance of John McMichael. Quis Separabit. Died At The Hands Of The Infamous IRA, December 22nd, 1987. Lest We Forget.* UDA, UFF, UVF, LPA.
>
> *In memory of Jack Duffin aged 66 years, Willie McManus aged 54 years, Christy Doherty aged 52 years, Peter Magee aged 18 years, James Kennedy aged 15 years. Murdered for their faith on 5th February 1992. Also in memory of other local people who have been murdered for their faith. Go ndéanaí Dia trócaire ar a n-anamacha.*
>
> *Restorick (Stephen) (Lance Bombardier 3RHA), killed at Bessbrook on February 12, his parents and brother wish to thank all who sent condolences. May his death be a catalyst for peace.*

Leonard has stated that the sheer volume of monuments erected in recent decades in Northern Ireland bear witness to the desire for public recognition of its losses through political violence. At the same time, partisan commemorations remain among the most divisive aspects of popular culture in Northern Ireland. This is equally true of monuments or parades. As Leonard concludes, a society has only matured when it can acknowledge that no side in any earlier conflict had a monopoly on heroism or honesty (Leonard, 1997).

Conclusion

Roadside memorials, whether personal or political, form part of the material culture of death and remembrance. Their location, outside of the conventional and acceptable contexts such as cemeteries or, in the case of political memorials, in public locations, disrupts attempts to contain these reminders of mortality.

The dead in Irish culture occupy, as Witoszek and Sheeran have argued, 'an almost tangible existence' (1994: 7), the various actions performed by the living, for the dead, as in the erection of roadside memorials can be considered an ongoing dialogue with the dead. In the Republic of Ireland the cultural practices that facilitate an ongoing connection are so embedded that they largely go unnoticed. However, within the island of Ireland, there are historically different, and frequently conflicting, cultural practices that are revealed in the responses to memorials in Northern Ireland. These responses reveal that even beyond death, there is a continuation of the conflict. If, as Leonard (1997) argues, a society has only matured when it can acknowledge that no side in any earlier conflict had a monopoly on heroism or honesty, the ongoing attacks and acts of vandalism on these memorials in Northern Ireland suggest that this point has not yet been reached. The materiality of these memorials, whether they are considered within a cultural tradition of roadside memorials or as war memorials, thus continues to disrupt, cause trouble and provides potent reminders of the conflict.

14
Geographies of the Spirit World

Douglas J. Davies

Most human cultures possess spaces for spirits understood as person-like powers capable of human interaction. While this may sound paradoxical, since spirits are typically deemed not to occupy a 'space' as do living persons, it becomes more intelligible once we recognise that 'life' itself is a mysterious entity difficult to 'place' or restrict to a body. This chapter will show how different societies classify their world and account for this interplay between life forces and the material nature of things is reflected in the way they make spaces for spirits.

Vitality, embodiment and mystery

The members of different cultures have conceptualised human 'life' or vitality in a variety of forms; for example, *samsara,* the Indian notion of the process of a transmigrating 'self' that inhabits a body during cycles of existence, and *psyche*, the classical Greek notion of *an* essence or spirit that enlivens the body and departs at death. One widespread symbol of such vitality is 'breath'; this ancient sign of 'life' is emphasised in the biblical creation myth where God breathes into human bodies formed from moist clay to animate them. These natural symbols of breath and breathing create movement, itself a prime sign of life, while the embodied stillness of their absence characterises death.

Throughout the world, millions of people believe that a soul quickens the embryo *in utero* and its departure causes the body's death; yet many others interpret these events biologically. The former often extend belief in a life-force to embrace the self's ongoing nature and destiny 'outside' its former body, while the latter see death as 'the end'. Although speculation over the evolution of religious ideas of deities from more ordinary concepts of souls, as propounded by E. B. Tylor (1832–1917), is properly criticised, the fact remains that his emphasis upon dreams and experiences of the dead reflect many popular attitudes today. After a century of disfavour, Tylor's (1958: 1–390) notion of animism still offers value as a description of the powers or energies attributed to certain objects or places by some New Age or Pagan groups. Their emphasis, however, is often upon life and not death, making them less death-conscious than most established religions. Established relatively recently, they have yet to cope with large numbers of dead people. Even so, such groups accord much significance to ancient ancestors and sites for engaging them.

Identity and the dead

At the heart of any discussion of spaces of spirits lies the identity, nature or ontological status of the living. Because personal identity is influenced by others, by lingering memories built into the life-narratives of living people, it is easy to explain how the dead 'appear' in dreams or are encountered by bereaved people.[1] Indeed, it is worth noting that 'identity' resembles what Augé described as one of those 'enormous' words that sometimes come in pairs and help define each other such as 'oblivion' and 'memory', 'life' and 'death'.[2] Philosophically speaking some terms can also be described as the 'imaginary' of each other, creating pairs that demand each other as, together, they make more than the sum of their parts, for example, 'life and death'.[3] Such partner concepts prompt people to think more concretely of 'imaginary spaces' of spirits, as when cemeteries or memorials speak as much of the 'dead' as of their still effectual spirits. In other words, to understand how the intangible spirit world is imagined, it is important to take account of how its counterpart world of the living is construed.

Human self-knowledge is grounded in the fact that we sense ourselves to be alive just as we sense others to be alive and, on encountering the death of others, we similarly sense that our vitality will also give way to mortality. This vitality–mortality 'imaginary' pair then helps define us as individuals. And it is on the 'spaces' associated with this intersection of vitality–mortality that this chapter focuses, spaces that educate by fostering self-knowledge of existence under the influence of human imagination. Philosophically speaking, it is precisely this creative imagination, especially under contexts of duress that 'brings the dead to life' and furnishes them with spaces of encounter with the living. So naturalistic an explanation is, however, inadequate for those who believe souls to have an independent existence after death, as in contemporary western Spiritualism. This perspective understands the dead to exist in the 'world of spirit', still bearing markers of identity, such as gender. Many different cultural conceptions of human and spirit lives exist, each with their own appropriately appointed spaces of spirits. In some African societies, for example, ancestors are viewed as male, reflecting a cultural view of old, living women as becoming less female as they age. 'Old women dry out and become "like men" before they die' is how Fardon describes Chamba attitudes to ancestors in their 'subterranean' world defined as 'cold' in comparison to the 'hot' world of the living.[4] In West Camaroon there is less concern with ancestors than 'the world of unborn children' described as 'a vast black cave, peopled by the spirits of children who wander around in pairs or groups looking for suitable parents'.[5] Issues of gender often relate a person's status when alive to that when dead. However, in one biblical statement, Jesus argues that when people rise from the dead they do not marry because they are 'like angels in heaven', the implication being that sexual division ceases in that domain.[6]

Paradigmatic spaces of spirits

Theoretically, we can describe such depictions as paradigmatic spaces of spirits, an expression derived from the wider notion of paradigmatic contexts to which numerous disciplines have contributed. For example, textual critic Alter developed the notion of a paradigmatic scene from the classical scholar Arend,[7] while anthropologist Needham[8] spoke of paradigm scenes as depicted situations enshrining central social values; these echo the sociological notion of ideal-type descriptions (Weber), albeit with a stronger narrative dimension. Again, anthropologist Maschio employed de Souza's notion of 'paradigm scenarios', not only to identify clusters of characteristic social values but also to include their associated emotions, an inclusion of importance for contexts of spirit contact.[9]

Allurement of place

Such paradigmatic spaces express what Jones described as the allurement of place which pervades what he identified as 'a ritual architectural event' in which a person's expectations merged with the opportunities of a particular, materially grounded locale.[10] The allurement of such events is often framed by historical-cultural traditions such as the Romantic movement of the eighteenth and the early nineteenth century whose notion of the 'sublime' described awe-inspiring places, albeit of a naturalistic kind. This 'secular' allure of place, complementing the human capacity to be thrilled, was taken further in Otto's theological notion of the sense of the holy with its *mysterium tremendum et fascinans* whose combined thrill, fear and excitement generates a kind of hopeful dread.[11] Underlying the possibility of ritual sites being places of spirit contact is the notion of the survival of the soul, whether derived from ancient Greek philosophy, Hinduism or Buddhist sources, or indigenous cultures. Certainly human experiences influence these philosophical perspectives and their preferred attitudes towards 'spirit', with some authors arguing that it is human experience of a mystical kind that generates the anticipation of a world beyond the body.[12]

Respectful sites

What is certain is that relationships between living people and spirits take many local forms. Among the Alaskan Tlingit, for example, where identity comprises multiple elements, there is a complex wealth of spaces of spirits with one being in the cemetery, another in the village of the dead and another in a reincarnated descendent.[13] Respect is shown to departed spirits through the potlatch feasts given to mark their passing into ancestor status. In this kinship-conscious society the space of the spirit encompasses a variety of forms, including living descendents dressed in the dead person's insignia. Another element of the dead person's identity also 'came back' to be 'reincarnated in a close matrilineal descendent' born soon after the death. They would be named at the celebratory event and greeted using their pre-death kinship terms. Even the physical likeness of the dead was thought to be re-created in the new-born child. Not to offer a celebratory feast was said to demean the deceased person amongst the spirits in the spirit-village of the afterlife.

Routes to evil

Alongside the positive emotions evoked by spirits encountered in sites of allurement or respect, spirits, believed to be misplaced or lost while journeying to the other world or malevolently inclined towards living kin because some ritual of respect has not been observed, can inspire fear. Further, a sense of lack of control over spirits or when the spirits display malevolent intent within the domestic sphere engenders anxieties. Accordingly, many traditional societies deploy funeral ritual to prevent the return of the dead, sometimes taking an indirect route to funeral sites or tying the corpse's limbs. Where malign spirit-influence occurs, as in haunting or poltergeists, help is routinely sought from ritual experts who engage with spirits to establish their reason for disturbing the living and to pacify them. Contemporary Japan takes spirit-presence in a different direction in rites in which certain used objects of everyday life, including writing brushes, tea utensils and sewing needles and sewing needles, are burnt in communal rites. Perhaps one might even say they were cremated at the end of their useful life. Albeit followed by a minority, this practice can still involve many thousands of people and for some is associated with attitudes towards 'the spirit of an object' which may hold 'malevolent power' or, 'in the case of dolls' even be the 'abode' of an unstable spirit.[14] The notion of evil spirits pervades locales in which fear is associated with the dead for, just as the living may harm others, so may the dead. Ancestors are at their best when 'blessing' their descendents but they may also curse them. This raises the issue of 'causes' that are imaginatively sought for the all too real ills befalling human beings. Until recent times, many illnesses have been blamed upon malign spirit forces and indeed today's world also hosts such ideas. In Hong Kong in the 1990s, for example, bullets used to execute criminals were sought and even dipped in the blood of execution to be used as a protection against evil spirits, the rationale being that in execution 'some sort of life force is released' and may help protect the holder of a charm associated with it.[15]

Welcomed homecoming

Driven by more positive emotions, on other occasions the spirits of the dead are welcomed back into society in a variety of contexts. As with the Alaskan Tlingit so with the Melanesian Rauto, whose memorial events involve someone dancing in the spirit-mask of one who has died. Through this paradigmatic scene of interaction between living and dead, the living kin gain a sense of the presence of the dead, as the mask 'brings back to life' a representation of the dead, with personal memories evoked once more. In traditional Melanesian men's houses skulls and other bones obtained in double-burial rites, which involve an initial disposal that precipitates the process of decomposition, followed by a second disposal, are retained to 'symbolize the power of the living as much as the power of the dead'. Trees, too, could be associated with the names of their deceased, ancestral planters that now 'embody the soulstuff of their namesakes'. Together such houses and trees constitute 'a single sociomythic space' and complement horticultural rites in which memories of former ritual specialists emotionally overwhelm the current practitioner who may even 'see' the faces of the dead in symbolic plants.[16] Maschio spoke of such experiences in terms of the Rauto notion of *makai* that embraces a 'plenitude of memory' and a kind of fullness of sadness that engenders a creative nostalgia. In this ritual event a person is allowed 'the feeling that comes from being composed of others, and of having another's lost presence once again touch one's own' (1994: 72). Through this emotional nexus bereaved people are provided with a cultural resource for developing their identity, a time of positive opportunity rather than retrospection.

This telling example shows how ideas of spirit can help define the nature of the living and their identity and how there can be 'space for spirit' not only in memorial halls, as in the Iban of Borneo who furnish small huts for spirits with 'miniature furnishings... sunhats, mats and baskets' that symbolise the status of the spirit whilst alive,[17] but also in the 'space' that is the *relationship* between the living and the dead. While this is easily seen in the explicit context of a séance where a medium brings the dead to a living relative, it is also evident when the living keep objects as reminders of the dead, or indeed as ways of retaining the dead in a more metaphorical sense, one that can nonetheless trigger an emotional awareness of links to another.[18] Those bereaved people who keep the rooms of the deceased 'just as they were' when the person was alive are, in some sense, maintaining a 'space for spirit', indeed, an untouched room offers a prime example of a paradigmatic scene of a relationship persisting beyond death, which some might see as pathological. Much the same can be said for many British people who keep the cremated remains of their partner or parent within their home. In western contexts those who sense a presence of the dead, perhaps a third of the population, may well not speak of it since it is not part of public discourse relating to life experience.[19] These ethnographic examples provide apt data that reinforce more recent therapeutic approaches to grief theory typified in the notion of 'continuing bonds' existing between the living and their dead as opposed to the earlier twentieth-century psychoanalytically driven idea of grief-work as a process of breaking life-time attachment to a deceased person (see Chapter 8).[20] The role of narrative-memory in the dynamic constitution of individual identity ensures a space for spirits in the ongoing life of surviving kith and kin, especially when intensified through anniversaries, photographic images of the dead and other vibrant moments of recall. Among the most grounded of continuing bonds is that which appears when a person sees in themselves a reflection of their own deceased kin.[21] Although to 'see' one's father in oneself is, essentially, a visual recall, it is also likely to evoke a biographical narrative.

A more widely acknowledged paradigmatic space for spirits is found in the Mexican Day of the Dead in which emotions associated with family unity and wider community solidarity in the face of death are evoked and given full rein.[22] Spirits are encouraged to return and join the living in celebratory commemoration focused on graves and involving carnival-like depictions of the dead as skeletons, skeleton-marked images, or even as confectionary. Emotions linked with community are facilitated by prevailing Roman Catholic religious beliefs in the existence of the soul after death and in the interplay between the living and the dead within the wide Communion of Saints of the Church.

Worshipful sites

For centuries Christian churches afforded a major link with departing or departed spirits through popular customs such as ringing bells, formal funeral rites, masses and prayers for the dead, with the idea of purgatory serving as the primary paradigmatic space for spirits in relation to the living. After an appropriate period in purgatory to cleanse souls from sin, they might, after resurrection in a 'spiritual body', arrive at their ultimate destiny – or destination – in the full presence of God in heaven. Prayers on their behalf, paid for by living kin or from their own estate, were believed to assist in this process and potentially reduce time required for purgation. Representations of purgatory helped foster a sense of this paradigmatic space of spirits and configured a map of existence which included a stage following earthly life and preceding a heavenly destination. As an imagined space, purgatory helped confer a sense of order and meaning on ritual surrounding and following death while also sustaining an extensive hierarchy and ecclesiastical domain grounded in the jurisdiction of the Catholic Church. King Henry VIII's break with Rome, aligned with the Reformation, changed this geography of spirits just as it altered the ecclesial-political map of Europe. For example, by 1560 in England long-standing theological arguments over purgatory, prayers for the dead, and paying money to gain indulgences to relieve souls from the penitential distress of purgatory came to a head. The English Reformation abolished monasteries and many places responsible for praying for the dead, while purgatory itself was defined out of existence as something fondly imagined and no longer to be entertained. That said, it was probably another 40 years before pockets of local practice were eliminated.[23] Though the demise of purgatory eliminated one of the most characteristic spaces of spirits within Christian societies, this did not mean that spirit beliefs ceased. Instead, they tended to find new sites for action as evil spirits or as ghosts.[24] Just as the Reformation led to a contest between Protestants and Catholics over places of worship so tensions emerged between old and new sites for spirits. By the time Loudon wrote his innovatory study on the planning of cemeteries in industrialised and urbanised Britain following a major visit to many European graveyards, he saw their 'main object' as fostering the safe decomposition of corpses and their 'secondary object' as 'the improvement of the moral sentiments and general taste of all classes... especially the great masses of society'.[25] These functions focused on the safety and the feelings of the living rather than on any socio-mythic approach to the places of the dead. The twentieth century heralded new circumstances and demands for new sites for the dead, especially soldiers. For example, following the Great War of 1914–1918, The War Graves Commission was charged with establishing cemeteries for the one million dead military personnel: some 2,000 war cemeteries were created in France and 500 in Belgium.[26] Visited by thousands of surviving relatives across Europe, these cemeteries also offer potential spaces for encounters with a spirit world.

With time, albeit not until the late decades of the twentieth century, the Church of England again allowed prayers for the dead, though not without some strong dissent from its more Protestant constituency. No longer aimed at seeking release from purgatory,[27] these newfound prayers were more pastoral in approach, acknowledging the need of living people to pray for those they loved and who may just have died, much as they might have prayed for them whilst still alive. Here, again, the paradigmatic space for spirits, as in prayers included in the Eucharist, becomes a shared space between the living and the dead: its moral bond and associated emotion is that of fondness rather than of fear as to the state of the dead. Yet, the space also encompasses emotions such as regret and guilt in a self-reflective acknowledgement of the believer's own moral imperfection and need for transformation before entering the fullness of the divine presence.

In more formal and technical theological terms, the regular worship of millions of Christians locates the idea of 'spirit' at the very centre of its theology by defining orthodox faith as belief in the Holy Trinity. This creedal formula of Father, Son and Holy Spirit took some 400 years to establish and has been debated ever since. Inheriting the motif of a divine breath or wind from Jewish scriptures, an entity overlying the chaotic waters prior to life's emergence (Genesis 1:2), and enlivening Adam, the man made from dust (Genesis 2:7), Christians could speak of that Spirit descending upon Jesus at his Baptism (Mark 1:10). This episode included a divine 'voice from heaven' and has often been used to validate the Trinity with the Father as the voice, the Son being baptised and the Spirit descending 'like a dove'.

Christianity's marked affinity for 'spirit-language' provided a valuable reference-point for 'experience' and for changes in a person's emotional sensations for the 'Spirit' is invisible yet ever potentially available. Significantly, believers were described as constituting a kind of temple for the Holy Spirit (1 Corinthians 6:19). The rise of the Charismatic Movement from the late 1960s amongst millions of people across the world reinforced belief that the Holy Spirit could and would become part of a human being's life as attested by the capacity to 'speak in tongues', to see visions, experience healings or to be 'slain in the Spirit', a phrase describing a sense of being overwhelmed by a power other than one's own. The logic of this Holy Spirit language easily embraced the cognate idea of evil-spirits to create new sites for spirits in a world readily entertaining notions of good and bad spirit-possession and of the possibility of exorcising the evil variety. Within more ordinary Christianity the Holy Spirit is associated with the rite of confirmation linked with baptism when a person becomes a Christian and, as such, is 'born of' the Spirit. Similarly, the Holy Spirit is invoked as part of the Eucharist rite in which bread and wine are blessed, with God being asked 'by the power of the Spirit' to let the elements become sacramental materials allowing believers to participate in the very body and blood of Jesus Christ who was sacrificed in crucifixion for the sins of the world.

Souls, spirits, Spiritualism

Like the concepts of human 'life' or vitality, introduced above, many Christian traditions speak of each individual as having or being a 'soul', a complex theological idea that is often understood by ordinary believers as some kind of an individual life-force that comes to the child in the womb, quickening it or giving it 'life', and departing the body at the time of death. Some modern Protestant theologians strongly oppose this notion to emphasise the miracle of the resurrection of the body as the basis of life after death. Whatever the particular emphasis, churches become major sites for the Holy Spirit's operation upon people, with religious groups often laying claim to particular conceptions of 'spirit', sometimes extending these into popular beliefs in malign spirit influence, a process which also occurs in some popular Islamic contexts.[28] Indeed, the wider world of popular religion often contains a complex mix of ways of bonding with the dead. Careful social scientific research on early twenty-first-century cemeteries in Britain, for example, revealed an intricate picture in which 'many…believe that the "spirit" of the deceased is present at the gravesite and that contact and communication is possible' (Francis *et al.*, 2005: 123) In their study these authors even report how some people seemed to suggest that there 'are three different "spirits" of the deceased', one going to heaven to rejoin deceased kin, another being at the grave and another as a 'presence' at home.[29] Great care is needed, then, when trying to classify ideas and bring order to the numerous and often partial ways of conceptualising spirits in any society. Boyer, for example, describes how the Fang of Cameroon talk about what might be called ghosts as some aspect of a dead person that, at first, seems to be part of 'a mass of agencies' of a malign nature that wander in the forest. A couple of months after the death, funeral rites transform this entity into an ancestor with a specific identity who becomes protective towards the living. But Boyer is keen to state that 'none of these notions should be taken too strictly as "doctrine" of the soul and its destiny…they do not constitute an integrated system'. He describes one place for spirits in which living people are initiated into their 'ancestor-cult'. This occurs at night in a building surrounded by a high fence. Ancestor skulls are placed in it and the novices take an hallucinogenic drink, enter a trance, and are shown the ancestors' skulls and given information on their genealogies.[30] In this state they go 'half-way' to the ancestors' village, entering a liminal space between the living and the dead where they converse with and receive messages from the ancestors. To go 'too far', however, would be to die. What we have called 'spaces for spirit' here shows a *distinction* between human domains and those of the dead and of the need for both contact and distance between them. Such spaces of spirits, places of allurement, possess just such a dual dynamic of attraction and repulsion.

Another distinctive site for spirits was created by the Spiritualist Churches that emerged from 1848 in America and gained some considerable following in many parts of the world as the nineteenth century closed. This movement understood the dead to exist 'in spirit' or in a spirit-world where they could contact the living and where living people could be in touch with their dead. In the United Kingdom, the intellectual-religious curiosity of the later nineteenth century, which also produced the Society for Psychical Research in 1882, was deeply reinforced by many people's experience of being bereaved of their sons and husbands during the First World War.[31] Given the silence of the churches over the immediate post-mortem state of the dead, Spiritualist Churches and independent mediums soon offered their own paradigmatic space for spirits in the context of a séance or 'sitting'. Whether in a specially designated Spiritualist church or in a private home, the medium acts as a channel for or bridge to the spirit world. He or she may remain awake and alert whilst seeking to connect the spirit and its earthly relative or may enter a trance state. Emotions of excitement and anticipation may mix with the underlying sense of loss and grief motivating bereaved people to seek Spiritualist help, especially in the weeks and months following a death. There may also be a sense of anxiety over those who have died, as survivors wonder if they are 'safe'. However, Spiritualist Churches, often situated in backstreets rather than alongside mainstream denominations on a town's main thoroughfare, remain small as far as regular membership is concerned, with the wider tidal flow of occasion seekers not remaining in the group on a long-term basis.

Sociomythic space

The idea of the sociomythic space refers to spaces in which living people may believe the dead reside and are enabled to relate to them, and which may be anticipated as part of their own destiny. Such spaces exist in complex philosophical forms as well as in popular ritual expressions as, for example, in traditional Tibetan groups influenced by the *Tibetan Book of the Dead*. Seldom are 'sociomythic spaces' more diverse and yet integrated than in the cluster of ritual practices that constitute Tibetan 'sky burial'. Here the corpse has its back broken at its home and is placed in a seated position before being taken to be cut into pieces and fed to vultures, all except a small piece of skull associated with the exit of the life-force. The third day after death is deemed the first of the deceased's journey in the liminal world of consciousness prior to reincarnation. A white, clay-covered jar is set up as an effigy of the deceased person's clothing and suspended in their former home. Temporarily, 'the consciousness of the deceased is believed to take residence in the jar, which is called a "spirit-place" (*thugs-sa*)'.[32] Subsequent rites seek to aid the journey of this consciousness through the in-between world of the *Bardo* state. These include another effigy of the dead person being placed before an altar where 'a block-print representing the dead is made on paper', and their 'consciousness' is transferred from jar to paper. The jar and effigy are destroyed and the paper undergoes a rite of burning that marks the forgiveness of the sins of the dead person and the freeing of their consciousness to undertake its transmigratory journey. The paper's ashes and the ground-up remains of the skull fragment are combined in small cones finally located at local sacred sites. Symbolic rituals such as these show how sociomythic spaces and spirit-places enable both people who are philosophically erudite and those with less educated backgrounds to have some sense of 'control of destiny' that serves 'as a powerful psychological tool to diminish the fear of death, and to make this ultimate event intelligible if not tolerable'.[33] Classically, in India, Banaras (or Varanasi) has been the key sociomythic space of spirits, a Holy city that carries an immense weight of religious significance in relation to the deities, life and death. It is the symbolic meeting point of this life and the onward journeying life of the inner animating life-force of individuals. To die here, to be cremated on the banks of the sacred river Ganges or to have one's cremated remains brought to the river is to attract great blessing on that journey. Those individuals who come here in their final days fast their way through a spiritually oriented dying process, one which allows their spirit to depart just as it arrived before birth.[34] In this sense a space for the dying becomes a space for the departing spirit.

Theories for spirit places

Some concepts useful for analysing the places of spirits have already been introduced: paradigmatic site, allurement of place and sociomythic space. To these may be added, in conclusion, three final interpretative locales: dreams, 'cognitive artefacts', and a further cognitive approach to perception of life, lifelessness and spirits.

In comparative terms, the obvious core locale of 'spirits' lies in the minds of living people, whether we speak psychologically or ethnographically, with dreams providing a valuable resource. Psychologically speaking, the dead constitute a kind of memory whose recall activates the emotional life of the present during sleep, often in ways that can persist during waking hours. While the capacity to recall ideas, images and events from the past and to experience them anew in the light of current life-experience is common to all human beings, just how this process of recall and experience is conceived depends upon the local culture's classification of the world. This is exemplified in Lienhardt's revealing account of traditional Dinka life in the Sudan. He noted, for example, how the verb 'to dream' meant 'to show to oneself', a process he related to how Dinka engaged with 'ghosts' through dreams and did not 'materialize' them as his western readers might. He had never heard of Dinka ghosts appearing in 'external physical space' (1961: 153–154). Their absence may be explained by the way that the Dinka took the initiative and made periodic offerings to the dead, thereby hoping to be left in peace since the dead only intrude when they have been ignored.

Clear evidence of the felt importance of not ignoring 'spirits' can be found in religious architecture across the world, from temples and churches to ancestral shrines and shamanic sites of spirit contact. Even Egypt's pyramids were transit points for launching the spirits of deceased pharaohs on their cosmic journey. One theoretical way of interpreting such spaces for spirits is as a 'cognitive artefact', defined as a 'physical structure which helps humans perform some cognitive operation'.[35] One such cognitive operation is the process of aligning oneself with the powers that influence life and are deemed to reside in agents such as ancestors or deities. These entities have often been elaborated into complex theologies that complement expressive ritual forms within architectural settings. Such spirits would not be the spirits they are without the places devised for their 'presence'. One interesting aspect of human religiosity lies in the way some societies populate their paradigmatic spaces for spirits with a high god, lower deities, intermediaries and ancestors. Other traditions, by contrast, strive to reduce this spirit population density as in the history of Israel when prophets fought for monotheism over and against polytheism. So, too, in Islam and its focus on a strict monotheism that even questions Christianity's strong monotheism that nonetheless presents itself as a Holy Trinity. From a critical perspective the key issue is that even monotheistic deities are a 'spirit' and, accordingly, their place of worship a 'space for spirit'. It remains the task of experts within these theological traditions to draw sharp distinctions between 'god', 'spirits' and 'ancestors'. Indeed, as some biblical scholars have shown, it is during periods of religious reformation, when leaders try to distinguish between a 'one god' doctrine and all other beliefs in local gods and spirits, that ideas of ancestors and of trying to contact the spirit of famous ancestors become forbidden (Blenkinsopp, 2004). In other words, spaces of spirits are not without their political as well as economic dimensions; certainly, they play significant roles in the dynamics of social status and social control in numerous societies while also being available for millions of people as one model for pondering the imponderables of human relations, memory and self-reflection.

Finally, from a different theoretical perspective, Boyer's cognitive theory of religion can be applied to spirits. This argues that 'religious representations contain both explicit counterintuitive principles... and implicit biological principles that constrain them'.[36] What then of spirits that are conceived as entities capable of making decisions and influencing people, despite their lack of any attributes of the normal life-form of a human body? While it is counter-intuitive to think that entities lacking a body can make decisions, there remains an intuitive human capacity to differentiate between animate and inanimate objects. This makes a corpse problematic and, driven by the idea that a dead body *once* had some life energy from a view 'grounded in an early sensitivity to the difference between' self-generated and non-self generated movement, there emerges the idea of a soul or spirit existing independently of the body.[37] Though speculative, this argument about human perceptions of the dead in relation to ideas of spirits is suggestive. Humans beings are characterised by movement while the dead are, essentially, still, a radical paradox that could generate the idea of a spirit in order to explain the difference. Boyer's conclusion resembles that of Tylor albeit by quite a different route since Tylor worked from assumptions of dreams and life forces, Boyer from the inherently human capacity to perceive the world in terms of animate and inanimate entities.

Notes

1. Many ethnographies and theoretical studies discuss 'the self'. Thomas Maschio (1994: 17–43) is especially valuable for analysing Melanesian notions of self in relation to issues of emotion, ethos and pathos.
2. Marc Augé ([1998] 2004: 13–14).
3. Cf. Jean Baudrillard ([1976] 1993: 133).
4. Richard Fardon (1990: xii, 183).
5. Robert Brain (1969: 216).
6. Mark 12: 25.
7. Robert Alter (1981: 50–51). Walter Arend (1933).
8. Rodney Needham (1981: 88–91).
9. Thomas Maschio (1994: 191). Maschio cites others including Rosaldo (1984) and de Souza (1990: 182).
10. Lindsay Jones (2000: 79).
11. Rudolph Otto ([1917] 1924). He called the experience of this tremendously attractive mystery the *sensus numinis*.
12. John S. Dunne (1965: 111–139).
13. Sergei Kan (1989: 182–183).
14. Angelika Kretschmer (2000: 402).
15. Virgil Kit-yiu Ho (2000: 156).
16. Thomas Maschio (1994: 182–185, 156, 161, respectively).
17. Clifford Sather (2003: 226–227). In William Wilder (ed.) (2003).
18. Margaret Gibson (2008).
19. Douglas Davies and Alastair Shaw (1995: 96) for UK. A. Grimby (1993: 75–76) for Sweden.

20. Cf. Klass, Silverman and Nickman (eds) (1996).
21. Katherine Young (2002).
22. Stanley Brandes (2006).
23. Peter Marshall (2002: 92–187).
24. R. C. Finucane (1982: 90–114).
25. J. C. Loudon (1843: 1).
26. *War Graves of the Empire* (1928: 1).
27. Notions of moral preparation of the dead for the vision of God still recur (cf. Douglas Davies, 2008: 86–94).
28. I. M. Lewis (1971).
29. Doris Francis, Leonie Kellaher, Georgina Neophytou (2005: 123).
30. Pascal Boyer (1994: 213).
31. Winter Jay (1995: 54–77).
32. This account and analysis taken from Claes Corlin (1988: 63–75).
33. Claes Corlin (1988: 74).
34. Christopher Justice (1997).
35. Edwin Hutchins (1994) cited by Roy D'Andrade (1995: 210).
36. Pascal Boyer (1994: 93, 119).
37. Ibid.: 107.

15
Recovering Presence

Jenny Hockey, Carol Komaromy and Kate Woodthorpe

The diversity of theoretical resources discussed in the introduction to this collection have enabled contributors to engage with embodied experiences of the imagined and experienced times and places associated with loss. In so doing, the collection offers a distinctive and nuanced explanation of how particular materialities can mobilise emotion, commitment and action. Theoretical debates about the relative persuasiveness of evidence for context-specific, rather than universal, responses to human mortality raise questions as to whether dying and loss are simply 'what we make of them', or whether weight – or agency – needs to he given to the materially grounded nature of all human life and death. We have contributed to these debates by making the perspective of material culture a starting point, and by exploring the embeddedness of human experience within time, space and place, within the context of objects and in relation to the human body.

This means that the data presented have *animated* death, dying and bereavement, opening up terrain eschewed by those who see such a topic as either depressing or pointless. Thus a material culture perspective has oriented us towards the embodied, materially grounded nature of human life. While 'the body' which sociologists, braving the critique of biological determinism, began to integrate into their work in the 1990s was treated as the *recipient* of human meaning-making (James and Hockey, 2007), the phenomenological concept of embodiment moves us on from this limited view into the experiential grounding of everyday life and, indeed, the life course. As such, embodiment places the sensory and emotional experiences of illness, caring, dying and loss at the heart of this enquiry. The lived body, actively engaging with its material environment, takes readers beyond the sanitised limits of social constructionism, on into the zones of bodily decay and emotional anguish.

While this volume's theoretical project has encompassed a range of disciplinary orientations, it is, nonetheless, grounded in a collection of 'experience-near' (Geertz, 1977) data. These demand that we share in the bodily, emotional and sensory experiences of particular individuals. Bronfen (1992) and Bauman (1992) have argued that members of contemporary western societies have systems of thinking and representation which partially deflect awareness of human mortality. Yet, precisely through their distinctiveness, the data presented in this collection render emotions such as fear, anger or grief, values such as dignity, respect or care, and practices such as cardio-pulmonary resuscitation (CPR), cremation or prayer, open to investigation and critique. As Geertz argued, it is through such data that the 'mega-concepts' of the social sciences – space, time, agency, identity, embodiment, sequestration – can be given the sort of sensible actuality that makes it possible to think not only realistically and concretely *about* them, but, what is more important, creatively and imaginatively *with* them (1975: 23).

In sum, this collection has offered a relatively extended theoretical discussion, moving through a range of empirical contexts, drawing on different perspectives to make sense of embodied experiences of death. The aim of this concluding chapter is to consider how we might 'think with' these perspectives, once fleshed out through data that allows them some kind of 'sensible actuality' (Geertz, 1975: 23). While Chapter 1 used the 'case of the body' to explore the value of different theoretical perspectives, this chapter puts space at the centre of its account of how the collection might enable us to think – and act – in relation to death.

Spatialities of dying, death, disposal and memorialisation

The work of Berger (1967), Bauman (1992), Seale (1998), Giddens (1991), and Mellor and Shilling (1993) have all contributed to what can be termed a 'sequestration thesis'. This highlights the temporal containment of death within an age-related cohort of people – those who have lived beyond the age of 65 (Holloway, 2007) – and its relative absence from the lives of children and younger adults; its spatial sequestration within hospitals, residential homes, funeral parlours and cemeteries; and its social seclusion among dying and bereaved people and the practitioners who work with and for them: hospice staff, coroners, pathologists, funeral directors, crematoria managers, the owners/managers of burial grounds, bereavement counsellors. Arguably, these temporal, spatial and social boundaries create an environment within which life and death come to be understood, within western societies, as a mutually defining, binary pair, twin domains of human experience that occupy an oppositional relationship with one another (Hockey, 1990).

This collection provides some support for these arguments and here we begin with the spatial organisation of living and dying, exploring the extent to which the collection evidences not only social commitment to, but also the achievement of a stable, spatialised life/death boundary. Set in the A & E department and the hospice, respectively, Chapters 2 and 3 examined the relationship between particular frameworks of ideas or values and their materialisation in the built environment. As they show, the social institutions of medicine, welfare and religion are manifested in the bricks-and-mortar of hospitals and hospices, in their location, internal design and equipment. This is not, however, a one-way journey from abstract concept to material expression. As Chapter 3 described, the concept of social welfare has a history spanning at least the nineteenth and twentieth centuries, one which found expression in all kinds of buildings – and in the experiences of those who have lived, died and worked in them. As Ingold (2000) would argue, understanding how 'places' such as hospitals and hospices come into being involves using a 'dwelling perspective'. Instead of a reading which reduces human experience to the imposition of an abstract blueprint upon an external environment, it is important to take account of historical processes of embodied engagement within sites of this kind – for example, the fear engendered by experiences of the workhouse. In other words, it is easier to understand this enmeshment of ideas and spatialised materialities by not thinking in terms of a divide between human beings and their environments crossed only by the imposition of cognitive systems *upon* that world. As Chapter 2 made plain, CPR, as an aspect of the hospital practice and experience of medicine, can evoke a sense of loneliness and desolation, not only among patients and their families, but also staff. When CPR is the final encounter between a sick person and their carers, we can interpret this potentially violent and impersonal transaction as a materially grounded experience with the capacity to breed resistance; for example, in the establishment of alternatives, such as the hospice movement and the development of palliative medicine.

If it is accepted that complex, lived experiences constitute 'place' *making* – as opposed to the spatialised implementation of abstract organising principles – then the notion of a fixed boundary between private and public space also becomes questionable. According to the sequestration thesis, the materialisation of a culturally specific value system within institutions represents the removal of death-related experiences from both the home and the local community (Elias, 1985). However, the data presented here show death to be one of the dimensions of everyday life that trouble fixed divisions between the public/institutional and the private/domestic. For example, Chapter 3 argued that while the institutions of medicine, welfare and religion have informed the establishment of palliative care, the experience of domestic life is also represented in hospice design. Whether palliative care is provided within the hospice itself, or the patient's home, the hospice movement forms part of a trend to return death to the domestic sphere from which it gradually came to be excluded during the nineteenth and twentieth centuries (Ariès, 1981; Adams, 1993; Gore, 2001). However, if the provision of care, which by the middle of the twentieth century had come to be seen as the province of professionals and institutions, has now begun to feature within domestic life, the presence of death within more public places (which might seem to sequester it from the spaces of living) can be paradoxically problematic. Thus, Chapter 3 describes strategies which serve to camouflage the death-related dimensions of dying places. For example, processional space, that is, the way in which visible movement through the hospice is orchestrated, tends not to include the dead body which instead leaves the building covertly. Similarly, the floral tributes from funerals which are often sent back to the hospice enter the building in the reverse direction to the dead body. Only when they have been re-arranged into domestic sprays and bouquets in the building's back regions can they be moved into its internal environment where they enhance its homely aspect. Along with the domestic sphere, other institutional cultures and aesthetics may provide ways of masking both the hospice and the A & E department's underlying orientation – even though *knowledge* of their associations with death is itself no secret. Thus, for example, traditional religious iconography may recreate the aesthetic of a place of worship, rather than illness in the hospice; the culture of medicine may transform the deathbeds of patients submitting to CPR into high-tech sites of curative practice where 'everything possible is being done'.

Chapter 4 developed these accounts of the complex spatial boundary between life and death, along with their focus on its mutability as an embodied experience. It addresses the paradoxical problem of containing death within a setting understood to frame the final years of the life course by exploring how space is organised to manage the ageing body, both architecturally and through embodied practice. While Chapter 2 described highly charged moments which marked the parting of the ways between living and dying for many patients, Chapter 4 takes the reader further into the embodied practices which, over time, are deployed in the service of sequestering death. However, just as A & E is not designed as the setting for a traditional death-bed, so residential care homes have to be adapted, through practice, to conform to requirements to confine death to particular spaces, times and periods of a resident's life course. That said, data from within this setting demonstrate that a commitment to containing 'dying' within a narrow margin at the very end of life is forever subverted by the ragged processes of deterioration which dog attempts to engineer 'rehabilitation'. Those apparently 'keeping going' as members of the 'living' category may die abruptly, so collapsing the intended spatial and temporal distance between the categories 'life' and 'death'. Thus the 'care' aspired to in policies geared towards humanising late-life ageing and overcoming the threats to social identity posed by institutional life, mutates into a system of control which categorises and contains, rather than responding to the individual requirements of people sharing the final period of their lives. Death, confined and made predictable, occurs only when the body produces irrefutable indicators of its final undoing.

Through the immediacy of case study data, Chapter 4 also demonstrated that embodied engagement with spaces of this kind has implications for the relationship between the body and identity. Who and what people are, the bodies they *are* or were, may well be severed from what is happening to the bodies they *have*. This dimension of spatialised, death-related experience – its implications for personhood and identity – is core to Chapters 5–7. Thus, just as the spatial boundaries through which death is sequestered prove complex when understood as forms of embodied place-making, so personhood and identity emerge as relational, negotiated concepts that refer to spatially located practices. Chapter 5 demonstrated that the apparently straightforward process of being born into a particular identity, such as 'son', 'granddaughter', 'brother' or 'girl', is in fact the material outcome of both spatialised practice and the agency of the bodies of both the child and its parents. Using an historical perspective, this chapter shows how the management of stillborn children's bodies, in space, can result in particular, contingent experiences of the child as either waste matter to be disposed of or a named family member to be memorialised. Set within the coroner's court, Chapter 6 similarly demonstrated the ways in which the body can be made to appear in particular forms, in particular places, different 'bodies' being mobilised in the accounts of pathologists, healthcare professionals and friends and family, their totality becoming the negotiated outcome of the coroner's decision-making process. Not only the identities of the judicial process participants and the deceased person, but also the kind of death that person died, thus emerge through relational transactions. This processual negotiation of identity through representations of the body therefore has theoretical resonance with the viewing of the body as a material entity. Chapter 7 provided detailed ethnographic evidence of how the making of places for viewing a dead body can rupture or precipitate processes of identification. The contingent nature of such processes was made persuasively evident through anthropological cross-cultural comparison; UK and US funeral directors' premises were the sites within and around which the bodies of the deceased person, their family and friends, and the ethnographer herself, together created an understanding of how death can trouble any easy relationship between the body we *have* and the body we *are*. In these data, bodies were both sequestered and made highly public through the conjoining of spaces, objects and people, so reflecting established practice in these two different social contexts.

Having argued that identity is relationally negotiated, this collection also takes account of the ways in which it is potentially disturbed in the separation of the social and object body at death (Hallam *et al.*, 1999). This becomes apparent when examining how the body is disposed of and the materialities drawn upon and generated through cremation and burial. Rather than face-to-face interaction as an important site of identification, the embodied presence of the dead person is eradicated through these forms of disposal. While burial, in a material sense, simply conceals their physical remains, Chapter 8 showed how embodied practices at the grave-site facilitate emotional engagement with the remembered social body of the dead person, rather than contemplation of their corpse, decaying in the earth. More profound 'concealment' is enabled through cremation, the damp, leaking materiality of the dead body being very rapidly transmuted into dry crumbs. Absence, a theme evoked in many of the chapters of this collection, is not, however, without its material dimensions. Using psychoanalytic terminology, Bronfen (1992) discusses the return of the 'repressed'. In other words, in the inevitably makeshift strategy of banishing death there is considerable scope for the return of those who have died, along with awareness of human mortality. Bronfen (1992) herself is concerned with the prominence of representations of dead women in European art and literature, a feature of patriarchal societies which, she argues, localises death's otherwise challenging universality at the site of the female 'other'.

Within the later chapters of this collection, concealment emerged as a complex phenomenon with unexpected outcomes. Chapter 9's focus on memorialisation within the City of London Cemetery showed how the practices of bereaved people not only evoked the presence of bodies lying under the headstones and memorial items, but also give material form to the persistence of relationships between people who are living and dead in the arrangement of flowers, windchimes, solar-powered croaking frogs, glass ornaments, laminated letters and toys on graves. The social identities of the dead are thus made accessible, as Chapter 10 also described in a particular natural burial ground where memorialising at the grave-site was left to personal choice, a choice that was liberally interpreted. However, both the cemetery and the natural burial ground are collective environments and the ways in which personal relationships are manifested in them have implications which extend beyond the individual. Thus, in the natural burial ground explored in Chapter 10, freedom of individual choice was eventually brought into line with practices at other such sites where memorialising and grave-marking tend to be restricted to organic matter such as leaves, twigs or the native species of trees and plants by owners and managers (Clayden *et al.*, 2009). In both the natural burial ground and also the cemetery, the practice of regulating what is placed upon a grave is undertaken primarily by staff, even though the removal of objects may sometimes be at the hands of bereaved people themselves. However, just as the materialities of dying spaces in the hospital, the hospice and the residential home reflect different sets of values and beliefs about ageing and dying, so the movements of objects into and within burial grounds evoke different forms of relatedness and different conceptions of the place of the dead within the future lives of the living. Data presented in both Chapters 8 and 10 provided striking examples of the *informal* regulation of relationships between the living and dead by visitors to the burial ground, whether bereaved or not, so highlighting the contradictions and ambivalences which characterise the evocation of intangible dimensions of everyday life: relationships and the imagined future.

Chapters 9 and 11 described the possibility of disposing of the dead outside institutional, bounded spaces. The separation of public and private space, and the mirroring of this in the relationship between death and life, is thus, again, confounded – and further evidence of the return of death to the domestic sphere is provided. Again, in the absence of the embodied presence of someone who has died, the materialities of particular environments of memory serve to make present the persistence of relationality between the living and the dead. Indeed, the deliberations about how a place for a deceased person might properly be made, described in Chapter 9's account of informal ash disposal, may often be precipitated by a sense of the absence of social connection within the cemetery. Thus a large and poorly maintained cemetery within which the bodies of the family or friends of the deceased person have not been interred or scattered may strike bereaved people as an anonymous 'space', rather than a 'place' constituted through memories of an embodied relationship shared with the dead person (Hockey *et al.*, 2010). Chapter 11's discussion of burial on private land or in the domestic garden made a powerful statement about the importance of space, place and materiality in generating relationships with deceased people. While the retention of human ashes within the home itself can inspire uneasiness among visitors to the bereaved person's home (Prendergast *et al.*, 2006), suggesting that the permanent fusion of living/dying space risks engendering some kind of ongoing, *embodied* relationship, the disposal of a whole body in land outwith the boundaries of institutional space can provoke feelings of anxiety that somehow both the dead and the living are putting themselves at risk; each one vulnerable to the presence of the other. Private land that contains one or more corpses thus conflates living and dying space in a potentially problematic way, 'messing up' prevailing classificatory systems (Douglas, 1966).

The sometimes secret disposal of ashes in ways and at sites known only to people who are immediately bereaved can result in an absence of visible evidence and an associated undermining of collective disposal and memorialising. Chapter 11 similarly discussed the invisibility of whole body disposal on private land, its provenance as a burial ground being imperceptible to those outside the immediately bereaved circle. These data complicate the boundary between public and private space, rendering its relationship with the separation of death from life ragged and unpredictable. As Chapter 11 showed in the case of private land, being unbounded, and therefore at least visually 'open' to the public, as was the case with an unfenced field containing a corpse, can provoke uneasiness.

Chapters 12 and 13 returned the reader to public space. They explored deaths which conform to the statistically normal event of death outside the home (Holloway, 2007) yet, more problematically, take place in unbounded, non-institutional public places. Chapter 12 discussed the impact of parental suicide during childhood, drawing upon two cases which involved a parent deliberately walking or driving some considerable distance from the domestic sphere. Dying on the road and in a wood, respectively, these individuals made places that acquired power and agency as ambiguous locations which disrupted dominant classificatory frameworks, death sites saturated with feelings of grief and abandonment. Lying beyond the domestic sphere, often cloaked in secrecy, and emotionally inaccessible to survivors, the dangerous energy of these polluted settings could be defused only by collapsing their distance from the here-and-now of survivors' everyday lives. As the images in Chapter 12 indicated, death is entirely invisible in these two settings. These are photographs of absence; only the surrounding text peoples them with the presence of those who died. Yet for someone like Will, the bereaved photographer who was visiting these traumatic sites for the first time, such absences can make the dead person more present than they may have been in life (Gibson, 2004).

Chapter 13 similarly described deaths which occurred in public space, but outwith any institutional boundary. The images of roadside memorials in this chapter nonetheless represented the *presence* of artefacts that mark the similarly violent demise of particular individuals. Unlike their invisible counterparts in public space, the materialities of roadside memorials in Northern Ireland are profoundly collective in that they speak to particular *collectivities*. As such, this example underscores arguments made throughout this volume about the potency of everyday materials and technologies: wood, stone, lettering. Set up at the roadside, these memorials are objects with agency. As such they participate within different networks of animate and inanimate entities; political affiliations render them the death sites of both enemies and victims, the two opposed categories condensed in the one memorial. Where memorials refer to bygone collectivities, however, those which have become politically neutral, the residual artefact in some sense 'disappears', forming an unremarked element among other roadside vegetation and artefacts.

Throughout this volume, tangible places, objects and persons have been core to the data presented. Creating boundaries which give form to unstable materialities emerged as a preoccupation within many of the chapters; these include the dying or dead body, the events leading up to or 'causing' a death, relations between the living and the dead, the memory and social identity of someone who has died. However, this collection does not uncritically reflect conceptions of death which are prevalent within western societies; approaches to death trained upon the body and its transitions between health and illness, animation and objectification. While Chapter 14 made the body its starting point, it concerned itself with the ways in which body-based experience and space, both imagined and experienced, informs human beings' conceptions of not only the dead but also the nature of life and its relationship with death. Breath, for example, provides an image of an enduring spirit or essence, something which animates the person in life and in death. Therefore, body-based and spatialised metaphors are important mechanisms for giving form to the otherwise intangible 'spaces of spirits'. These metaphors, drawn from the familiar domains of embodied engagement with material spaces, do not simply give imagined shape to the afterworlds of human life. Rather, they take substance within ritual encounters, the materialisation of root metaphors being a characteristic of ritual practice (Turner, 1974). As the chapter richly illustrated, such sites take a variety of forms and, importantly, are places made with a variety of purposes; sometimes to welcome and engage with the dead, sometimes to establish and shore up boundaries between the living and the unwelcome dead.

Sequestration revisited

This chapter began by citing the considerable evidence provided in the collection for a sequestration thesis. Where it departs from any straightforward premise, however, is in relation to the notion that somehow death is intrinsically and universally fearful, a premise which is at times too readily used to 'explain' the sequestration, concealment and control of human mortality. In other words, the sequestration thesis is built upon the assumption that death everywhere and for all time has threatened human beings; all that needs explaining are the reasons why a particular set of mechanisms emerged and remain prevalent in modern western societies. Bauman explains the intrinsic 'problem' of death as follows:

> Death itself cannot be perceived; still less visualised or 'represented'. There is no 'something' which is death; nothing in which the stretched intention of the subject struggling for perception would rest, where it could cast its anchor. Death is an absolute nothing and 'absolute nothing' makes no sense.
>
> (1992: 2)

From this perspective, the human tragedy is to be endowed with memory and imagination which not only empower but also shadow self-awareness with knowledge of life's ephemerality. In Bauman's view, human beings make overdetermined bids to substitute some form of permanence for their own impermanence – children, wealth and memorable achievements – when faced with the unforgettable knowledge of their own mortality. Willmott (2000), however, is among those who criticise this argument. He frames his challenge in terms of the specificity of western values and meanings, arguing that these have shaped theorising around death and dying when, in fact, they should have been its focus. Against the views expressed by authors such as Bauman (1992), Willmott says that fear of death could be understood as:

> [t]he projected loss of what we (commonsensically) know, or believe, ourselves to be: the socially constructed sense of self-identity that arises simultaneously with participation in modern institutions.
> (2000: 656)

In his view, 'ontological and existential anxieties . . . are not universal or endemic to the human condition but, instead, are expressions of the socially organised privileging of a separation between wo/man and world' (2000: 657). In seeking to contain or bring to order our mortality, Willmott argues that we are striving to conserve the stability of bureaucratised, institutional life in contemporary western societies. This is something that Berger (1967), Bauman (1992) and Seale (1998), by contrast, see as the *consequence* of an innate fear of death; put at its most extreme, culture is that which we fling together to stave off the fragility of our embodied lives. Death, in this view, is conceptualised as a rupture; 'life' is abbreviated, often unceremoniously, by death. However, the work of Ellis (2009) among a sample of families where a member has a life-limiting illness suggests that this view obscures the everyday nature of some dying trajectories, and neglects the prevalence of an *intersection* of living and dying as a way of life for some families. Seremetakis (1991), too, argues that the notion of death as rupture, a marginal or boundary event which culture seeks to repair, neglects experiences of death as a site from which to *engage*, critically, with culture.

In the work presented in this book we have shown how the marginalising or sequestration of death can render it all the more present within everyday life. Indeed, the negotiated, unstable processes of only partial containment and avoidance described in many chapters can be seen to offer moments of resistance that spawn other less indirect encounters with loss. As we have shown, these more direct encounters may be discovered within the process of dying, the loss of a child at birth, the viewing of a body or the tending of a grave, the retention of ashes, the 'bringing home' of dying and disposal, the closing of potent distances between traumatic childhoods and more empowered adulthood, the spatialised encounter with the terrain of spirits. The idea that 'fearful' death is inevitably a focus for distancing and control is thus belied by the accounts in this collection which demonstrate the negotiated and indeed partial nature of any such distancing, the materialisation of presence that paradoxically results from its sequestration.

Bibliography

Adams, S. (1993) 'A gendered history of the social management of death and dying in Foleshill, Coventry, during the inter-war years', in D. Clark (ed.) *The Sociology of Death* (Oxford: Blackwell Publishers/The Sociological Review).

Age Concern (2008) *Our Statement on the 60th Anniversary of the NHS* (www.ageconcern.org.uk, 2 February).

Ahmad, S. and O'Mahony, M. S. (2005) 'Where older people die: a retrospective population-based study', *Quarterly Journal of Medicine*, **98**: 865–870.

Alessi, J. (2006) *Descansos: The Sacred Landscapes of New Mexico* (New Mexico: Fresco Fine Art Publishers).

Alter, R. (1981) *The Art of Biblical Narrative* (New York: Basic). *Anthropological Society (N.S.)*, **12**: 881–898.

Anderson, R. (2009) 'Blithe Spirit', Guardian newspaper, 3 Jan.

Appadurai, A. (ed.) (1986) 'Introduction: commodities and the politics of value', *The Social Life of Things: Commodities in Cultural Perspective* (Cambridge: Cambridge University Press).

Arber, R. (2000) 'Disposal of cremated remains: a European perspective', Unpublished report to the International Cremation Federation, by Roger Arber, Secretary-General.

Arend, W. (1933) *Die Typischen Scenen bei Homer* (Berlin: Weidmannsche Buchhandlung).

Ariès, P. (1962) *Centuries of Childhood* (London: Cape).

Ariès, P. (1981) *The Hour of our Death* (London: Penguin).

Armour, M. (2003) 'Meaning making in the aftermath of homicide', *Death Studies*, **27**: 519–540.

Armstrong, D. (2002) 'Conceptualising the patient', *Journal of Health Services Research Policy*, **7** (4): 245–247.

Árnason, A. (2000) 'Biography, bereavement, story', *Mortality*, **5** (2): 189–204.

Arnason, A. (2009) 'Death in Iceland, hidden and revealed', in S. Earle, C. Bartholomew, and C. Komaromy (eds) *Making Sense of Death, Dying and Bereavement: An Anthology* (London: Sage/The Open University).

Arnold, C. (2006) *Necropolis. London and Its Dead* (London: Simon and Schuster).

Ash, J. (1996) 'Memory and objects', in P. Kirkham (ed.) *The Gendered Object* (Manchester: Manchester University Press).

Augé, Marc ([1998] 2004) *Oblivion* (Minneapolis, London: University of Minnesota Press).

Bachelor, P. (2004) *Sorrow and Solace: The Social World of the Cemetery* (Amityville, NY: Baywood Publishing).

Baker, M. (2009) '170 outraged people sign cemetery petition', *The Evening Telegraph*, peterboroughtoday.co.uk, 23rd March, available online at http://www.peterboroughtoday.co.uk/news/170-outraged-people-sign-cemetery.5098845.jp [accessed 30 September 2009].

Barker, D. (2001) 'Colorado court rules roadside cross "litter" not venerated object', *Freethought Today*, www.ffrf.org/fttoday/2001/junejuly01/colorado.html [accessed 15 January 2010].
Barlow, C. A. and Coleman, H. (2003) 'The healing alliance: how families use social support after a suicide'. *Omega*, **47** (3): 187–201.
Barrett, T. W. and Scott, T. B. (1990) 'Suicide bereavement and recovery patterns compared with nonsuicide bereavement patterns', *Suicide and Life Threatening Behaviour*, **20** (1): 1–15.
Baudrillard, J. ([1976] 1993) *Symbolic Exchange and Death* (London: Sage).
Bauman, Z. (1992) *Mortality, Immortality and Other Life Strategies* (Cambridge: Polity Press).
BBC (2003) 'Cemetery visitors to be escorted', http://news.bbc.co.uk/1/hi/scotland/3084200.stm [accessed 30 September 2009].
Becker, S. H. and Knudson, R. M. (2003) 'Visions of the dead: imagination and mourning', *Death Studies*, **27**: 691–716.
Bender, B. (1993) 'Introduction. Landscape – meaning and action', in B. Bender (ed.) *Landscape: Politics and Perspectives* (Oxford: Berg).
Bender, B. and Winer, M. (2001) *Contested Landscapes: Movement, Exile and Place.* (Oxford: Berg).
Bennett, G. and Bennett, K. M. (2000) 'The presence of the dead: an empirical study', *Mortality*, **5** (2): 139–157.
Berg, M. and Bowker, G. (1997) 'The multiple bodies of the medical record: towards a sociology of an artifact', *Sociological Quarterly*, **38**: 513–525.
Berger, P. (1967) *The Sacred Canopy. Elements of a Sociological Theory of Religion* (New York: Anchor Books).
Biddle, L. (2003) 'Public hazards or private tragedies? An exploratory study of the effect of coroners' procedures in those bereaved by suicide', *Social Science & Medicine*, **56** (5): 1033–1045.
Blackburne, F. (1780) *Memoirs of Thomas Hollis* (London: J. Nichols).
Blenkinsopp, J. (2004) *Treasures Old and New: Essays in the Theology of the Pentateuch* (Grand Rapids: Eerdmans).
Bleyen, J. (2005) *De dood in Vlaanderen: opvattingen en praktijken na 1950* (Leuven: Davidsfonds Uitgeverij).
Bleyen, J. (2009a) 'Appropriate death', in C. Bryant and D. Peck (eds) *Encyclopaedia of Death and the Human Experience* (Thousand Oaks: Sage).
Bleyen, J. (2009b) 'Defining and conceptualizing death', in C. Bryant and D. Peck (eds) *Encyclopaedia of Death and the Human Experience* (Thousand Oaks: Sage).
Bloch, M. and Parry, J. P. (1982) *Death and the Regeneration of Life* (Cambridge: Cambridge University Press).
Booth, T. (1985) *Home Truths: Old People's Homes and the Outcome of Care*, (Aldershot Hants: Gower Publishing Co.).
Bourdieu, P. (1977) *Outline of a Theory of Practice* (Cambridge: Cambridge University Press).
Bowman, L. (1959) *The American Funeral: A Study in Grief, Extravagance, and Sublimity* (Washington, DC: Public Affairs Press).
Boyer, P. (1994) *The Naturalness of Religious Ideas* (Berkeley: University of California Press).
Bradbury, M. (1996) 'Representations of "good" and "bad" death among deathworkers and the bereaved', in G. Howarth and P. C. Jupp (eds) *Contemporary Issues in the Sociology of Death, Dying and Disposal* (Basingstoke: Macmillan).

Bradbury, M. (2001) 'Forget me not: memorialisation in cemeteries and crematoria', in J. Hockey, J. Katz, and N. Small (eds) *Grief, Mourning and Death Ritual* (Buckinghamshire: Open University Press).

Bradfield, J. B. (1993) *Green Burial: the d-i-y guide to law and practice* (London: Natural Death Centre).

Brain, R. (1969) 'Friends and twins in Bangwa', in M. Douglas and P. M. Kaberry (eds) *Man in Africa* (London: Tavistock Publications).

Brandes, S. (2006) *Skulls to the Living, Bread to the Dead, The Day of the Dead in Mexico and Beyond* (Oxford: Blackwell).

Bremborg, A. (2006) 'Professionalization without dead bodies: the case of Swedish funeral directors', *Mortality*, **11** (3): 270–285.

Bridging Work Group (2009) 'Bridging the gap between research and practice in bereavement', in S. Earle, C. Komaromy, and C. Bartholomew (eds) *Death and Dying: A Reader* (London: Sage).

Bronfen, E. (1992) *Over Her Dead Body. Death, Femininity and the Aesthetic* (Manchester: Manchester University Press).

Brooks, C. (1989) *Mortal Remains: The History and Present State of the Victorian and Edwardian Cemetery*. Exeter (Devon: Wheaton Publishers Ltd).

Broom, D. H. and Woodward, R. V. (1996) 'Medicalisation reconsidered: toward a collaborative approach to care', *Sociology of Health and Illness*, **18** (3): 357–378.

Brown, C. G. (2001) *The Death of Christian Britain: Understanding Secularisation, 1800–2000* (London: Routledge).

Browning, B. (2004) 'Playing dead: exploring the recreational use of cemeteries'. Paper given at Cemetery Research Group Colloquium (University of York, May).

Brummell, S. (1998) 'Resuscitation in the A&E Department: can concepts of death aid decision making?' *Accident and Emergency Nursing*, **6**: 75–81.

Buchli, V. (ed.) (2002) 'Architecture and the domestic sphere', *The Material Culture Reader* (Oxford: Berg).

Buchsbaum, C. C. (1996) 'Remembering a parent who has died: a developmental perspective', in D. Klass, P. R. Silverman, and S. L. Nickman (eds) *Continuing Bonds* (London: Taylor and Francis).

Burke, P. (2005) 'Performing history: the importance of occasions', *Rethinking History*, **9** (1): 35–52.

Seremetakis, C. (ed.) (1994) *The Senses Still: Perception of Memory as Material Culture in Modernity* (Chicago: Chicago University Press).

Carvel, J. (2007) *No Dignity for Older Patients on NHS Wards* (London: *The Guardian*, 27 September).

Carvel, J. (2008) *Patients not NHS Staff's Top Priority* (London: *The Guardian*, 9 April).

Casper, M. J. and Moore, L. J. (2009) *Missing Bodies: The Politics of Visibility* (London: New York University Press).

Catedra, M. (1992) *This World, Other Worlds: Sickness, Suicide, Death and the Afterlife among the Vaquieros de Alzade of Spain* (Chicago: University of Chicago Press).

Cathedra, M. (2004) 'Kinds of death and the house', in A. Robben (ed.) *Death, Mourning and Burial* (Oxford: Blackwell Publishing).

Cecil, R. (ed.) (1996) *The Anthropology of Pregnancy Loss: Comparative Studies in Miscarriage, Stillbirth and Neonatal Death* (Oxford: Berg).

Cheal, D. (1988) 'The postmodern origin of ritual', *Journal for the Theory of Social Behaviour*, **18** (3): 269–290.

Churchill, A. and Tay, C. (2008) 'An assessment of roadside memorial policy and road safety', *Canadian Journal of Transportation*, **2** (1): 2–12.

City of London Cemetery and Crematorium (2005) *Conservation Management Plan*, prepared by D. Lambert, J. Rugg, L. Kellaher, S. Crouch, J. Woodcock of Richard Griffiths Architects, and R. Thorne and J. Lineham of Alan Baxter and Associates London: City of London.

Clark, D. (2002) 'Between hope and acceptance: the medicalisation of dying', *British Medical Journal*, **323**: 905–907.

Clark, J. and Franzmann, M. (2002) 'Roadside memorials', *Pointers Journal of the Christian Research Association* (September 21–23) (Nunawading, VIC, Australia: Christian Research Association).

Clarke, J. and Cheshire, A. (2003) 'RIP by the roadside: a comparative study of roadside memorials in New South Wales, Australia and Texas, United States', *Omega*, **48** (3): 203–222.

Clayden, A. (2003) 'Woodland burial', *Landscape Design*, **322**: 22–25.

Clayden, A. (2004) 'Natural burial, British style', *Landscape Architecture*, **94** (5): 68–77.

Clayden, A. and Dixon, K. (2007) 'Woodland burial: memorial arboretum versus natural native woodland?' *Mortality*, **12** (3): 240–260.

Clayden, A., Hockey, J., and Green, T. (2009) 'Natural Burial and the Materialisation of Absence', Paper presented at the 9th International Social Context of Death, Dying and Disposal Conference (University of Durham, September 9–12).

Cohen, A. (1985) *The Symbolic Construction of Community* (London: Routledge).

Collins, C. and Rhine, C. (2003) 'Roadside memorials', *Omega* **47** (3): 221–244.

Cooper, S., Janghorbani, M., and Cooper, G. (2006) 'A decade of in-hospital resuscitation: outcomes and prediction of survival?' *Resuscitation*, **68** (2): 231–237.

Corlin, C. (1988) 'The journey through the Bardo: notes on the symbolism of Tibetan mortuary rites and the Tibetan book of the dead', in S. Cederroth, C. Corlin, and J. Lindstrom (eds) *On the Meaning of Death (Uppsala University), Studies in Cultural Anthropology*, **8**: 63–75.

Corr, C. A. (2002) 'Revisiting the concept of disenfranchised in grief', in K. Doka (ed.) *Disenfranchised Grief. New Directions, Challenges and Strategies for Practice* (Illinois: Research Press).

Crace, J. (1999) *Being Dead* (London: Penguin Books Ltd).

Croft, J. (2007) 'I live in a National Trust House' (*Guardian Weekend*, 12 May 1997).

Csordas, T. (1994) *Embodiment and Experience* (Cambridge: Cambridge University Press).

Csordas, T. (2002) *Body/Meaning/Healing* (Basingstoke: Palgrave Macmillan).

Curl, J. S. (1980) *A Celebration of Death: A Guide to Some of the Buildings, Monuments, and Settings of Funerary Architecture in the Western European Tradition* (London: Constable).

Curl, J. S. (1993) *A Celebration of Death*, 2nd edn (London: Batsford).

Currier, J. M., Holland, J. M., and Neimeyer, R. A. (2006) 'Sense-making, grief and the experience of violent loss: toward a meditational model', *Death Studies*, **30**: 403–428.

D'Andrade, R. (1995) *The Development of Cognitive Anthropology* (Cambridge: Cambridge University Press).
Danforth, L. M. (1982) *The Death Rituals of Rural Greece* (Princeton: Princeton University Press).
Davey, B. (2001) 'Do-not-resuscitate decisions: too many, too few, too late?' *Mortality*, **6** (3): 247–264.
Davidson, J., Bondi, L., and Smith, M. (eds) (2005) *Emotional Geographies* (Aldershot: Ashgate).
Davies, D. J. (2002) *Death, Ritual and Belief: The Rhetoric of Funerary Rites*, 2nd edn (London: Continuum).
Davies, D. J. (2005) *A Brief History of Death* (Oxford: Blackwell Publishers).
Davies, D. J. (2008) *The Theology of Death* (London: Continuum).
Davies, D. J. and Guest, M. J. (1999) 'Disposal of cremated remains', *Pharos International*, **65**: 26–30.
Davies, D. J. and Mates, L. H. (2005) *Encyclopaedia of Cremation* (Aldershot: Ashgate).
Davies, D. J. and Shaw A. (1995) *The Reuse of Graves: A Report of Popular British Attitudes* (Crayford: Shaw and Son).
Davis, G., Lindsay, R., Seabourne, G., and Griffiths-Baker, J. (2002) *Experiencing Inquests, Home Office Research Study 241* (London: Home Office).
Degrémont, N. (1998) 'Palliative care and architecture: from hospital to people', *European Journal of Palliative Care*, **5** (4): 127–129.
Del Vecchio Good, M. J. (1995) 'Cultural studies of biomedicine: an agenda for research', *Social Science and Medicine*, **4**: 461–473.
Demi, A. S. and Howell, C. (1991) 'Hiding and healing: resolving the suicide of a parent or sibling', *Archives of Psychiatric Nursing*, December **6**: 350–356.
Department of Health (2008) *End of Life Care Strategy* (London: DH).
Doka, K. (ed.) (2002). 'Introduction', *Disenfranchised Grief. New Directions, Challenges and Strategies for Practice* (Illinois: Research Press).
Dorries, C. P. (2004) *Coroners' Courts: A Guide to Law and Practice* (Oxford: Oxford University Press).
Douglas, M. (1966) *Purity and Danger. An Analysis of Concepts of Pollution and Taboo* (London: Penguin Books Ltd).
Douglas, M. (1966) *Purity and Danger: An Analysis of Concepts of Pollution and Taboo* (London: Routledge and Kegan Paul).
Draper, J. (1967) *The Funeral Elegy and the Rise of English Romanticism* (London: Frank Cass).
Dunk, J. and Rugg, J. (1994) *The Management of Old Cemetery Land: Now and The Future: A Report of the York University Cemetery Research Group* (Crayford: Shaw and Sons).
Dunne, J. S. (1965) *The City of the Gods, A Study in Myth and Mortality* (London: Sheldon Press).
Easthope, L. (2009) 'Returning property after death and disaster', in S. Earle, C. Bartholomew, and C. Komaromy (eds) *Making Sense of Death, Dying and Bereavement: An Anthology* (London: Sage in association with the Open University).
Edensor, T. and Millington, S. (2009) 'Illuminations, class identities and the contested landscapes of Christmas', *Sociology*, **43** (1): 103–121.
Elias, N. (1978) *The Civilizing Process: Volume 1, The History of Manners* (New York: Urizen Books).

Elias, N. (1985) *The Loneliness of Dying* (Oxford: Basil Blackwell).
Ellis, J. (2009) *Challenging the Psychological Concept of Denial: And Other Emotional Assumptions Related to Death and Dying,* Paper presented at the 9th International Social Context of Death, Dying and Disposal Conference (University of Durham, September 9–12).
Emke, I. (2002) 'Why the sad face? Secularization and the changing function of funerals in Newfoundland', *Mortality*, **7** (3): 269–284.
English Heritage (2007) *Paradise Preserved: An Introduction to the Assessment, Evaluation, Conservation and Management of Historic Cemeteries* (Published with advice and funding from Natural England, London: English Heritage).
Everett, H. (2002) *Roadside Crosses in Contemporary Memorial Culture* (Austin: Texas University Press).
Excell, G. (2004) *Roadside Memorials in the UK,* Unpublished MA thesis (University of Reading).
Fardon, R. (1990) *Between God, the Dead and the Wild* (Edinburgh: Edinburgh University Press for the International African Institute, London).
Featherstone, M. (1991) 'The body in consumer culture', in M. Featherstone, M. Hepworth, and B. Turner (eds) *The Body: Social Process and Cultural Theory* (London: Sage Publications).
Fernandez, J. W. (1972) 'Persuasions and performances: of the beast in every body. And the metaphors of everyman', Daedalus, *Journal of American Academy of Arts and Sciences, Myth, Symbol and Culture* **101** (1): 39–60.
Fielden, J. M. (2003) 'Grief as a transformative experience: weaving through different lifeworlds after a loved one has completed suicide', *International Journal of Mental Health Nursing,* **12**: 74–85.
Fincham, B., Scourfield, J., and Langer, S. (2008) 'The impact of working with disturbing secondary data: reading suicide files in a coroner's office', *Qualitative Health Research,* **18**: 853–862.
Finucane, R. C. (1982) *Appearances of the Dead, A Cultural History of Ghosts* (London: Junction Books Ltd).
Foucault, M. (1973) *The Birth of the Clinic: An Archaeology of Medical Perception* (London: Tavistock).
Foucault, M. (1977) *Discipline and Punish: The Birth of the Prison* (London: Penguin).
Foucault, M. (1984) 'Des Espace Autres', *Architecture /Mouvement/ Continuité,* **5**: 46–49.
Fox, N. (1997) 'Space, speciality and surgery: circuits of hygiene in the operating theatre,' *Social Science and Medicine,* **45** (5): 649–657.
Francis, D. (1997) *A Cemetery for Posterity: The Conservation of the Landscape of the City of London Cemetery,* unpublished MA dissertation (London: Architectural Association).
Francis, D., Kellaher, L., and Neophytou, G. (2001) 'The cemetery: the evidence of continuing bonds', in J. Hockey, J. Katz, and N. Small (eds) *Grief, Mourning and Death Ritual* (Buckingham and Philadelphia: Open University Press).
Francis, D., Kellaher, L., and Neophytou, G. (2005) *The Secret Cemetery* (Oxford: Berg).
Frank, A. W. (1991) 'For a sociology of the body: an analytical review', in M. Featherstone, M. Hepworth and B. S. Turner (eds) *The Body: Social Process and Cultural Theory* (London: Sage).

Frosch, D. (2008) 'Wyoming finds a place for crash-site memorials' (*International Herald Tribune*, October 26).

Gadher, D. (2004) 'Councils ban roadside shrines to crash victims' (*The Sunday Times*, October 24).

Garattini, C. (2007) 'Creating memories: material culture and infantile death in contemporary Ireland', *Mortality*, **12** (2): 193–206.

Garces-Foley, K. and Holcomb, J. (2006) 'Contemporary American funerals: personalizing tradition', in K. Garces-Foley (ed.) *Death and Religion in a Changing World* (New York: M.E. Sharpe).

Geertz, C. (1975) *The Interpretation of Culture* (London: Hutchinson).

Geertz, C. (1977) 'From the native's point of view', in J. L. Dolgin, D. S. Kemnitzer and D. M. Schneider (eds) *Symbolic Anthropology* (New York: Columbia University Press).

Gell, A. (1998) *Art and Agency: An Anthropological Theory* (Oxford: Clarendon).

Gibson, J. J. (1979) *The Ecological Approach to Visual Perception* (Boston, MA: Houghton Miffin).

Gibson, M. (2004) 'Melancholy objects', *Mortality*, **9** (4): 285–299.

Gibson, M. (2008) *Objects of the Dead, Mourning and Memory in Everyday Life* (Victoria: Melbourne University Press).

Giddens, A. (1991) *Modernity and Self-Identity: Self and Society in the Late Modern Age* (Cambridge: Polity Press).

Gilbert, S. (2006) *Death's Door: Modern Dying and the Way We Grieve* (London: W. W. Norton).

Gittings, C. (1999) 'Sacred and secular: 1558–1660', in P. C. Jupp and C. Gittings (eds) *Death in England, An Illustrated History* (Manchester: Manchester University Press).

Gittings, C. (2007) 'Eccentric or enlightened? Unusual burial and commemoration in England, 1689–1823', *Mortality*, **12** (4): 321–349.

Gittings, C. and Walter, T. (2009) *Garden Burial: Contemporary and Historical Perspectives* (unpublished paper).

Goffman, E. (1961) *Asylums* (London: Penguin Books).

Goffman, E. (1963a) *Stigma: Notes on the Management of Spoiled Identity* (Englewood Cliffs, NJ: Prentice Hall).

Goffman, E. (1963b) *The Presentation of Self in Everyday Life* (London: Penguin Books).

Goffman, E. (1968) *Asylums: Essays on the Social Situation of Mental Patients and other Inmates* (Harmondsworth: Penguin).

Gore, P. (2001) 'Funeral ritual past and present', J. Hockey, J. Katz, and N. Small (eds) *Grief, Mourning and Death Ritual* (Buckingham: Open University Press).

Gore, P. (2007) *Embalming and the Social Construction of the Corpse in Contemporary England*, Unpublished PhD thesis (Brunel University).

Gorer, G. (1965) *Death, Grief and Mourning* (New York: Routledge).

Gosling, P. (1993) *The Archaeological Inventory of co Galway. Volume 1: West Galway* (Dublin: The Stationary Office).

Grainger, H. J. (2005) *Death Redesigned, British Crematoria: History, Architecture and Landscape* (Reading: Spire Books).

Graves-Brown, P. (ed.) (2000) 'Introduction', *Matter, Materiality and Modern Culture* (London: Routledge).

Grider, S. (2001) 'Spontaneous Shrines: a modern response to Tragedy and Disaster', *New Directions in Folklore*, **5**: 12–17.

Grider, S. (2005) 'Spontaneous shrines and public memorialization', in K. Garces-Foley (ed.) *Death and Religion in a Changing World* (Armonk, NY: M.E. Sharpe).

Grider, S. (2006) 'Descansos: the sacred landscapes of New Mexico', in J. Alessi (ed.) (2006) *Descansos: The Sacred Landscapes of New Mexico* (New Mexico: Fresco Fine Art Publishers).

Grimby, A. (1993) 'Bereavement among elderly people: grief reactions, post-bereavement hallucinations and quality of life', *Aspects of Quality of Life in Old Age, Acta Psychiatr Scand*, **87**: 72–80.

Grimes, R. L. (2000) *Deeply into the Bone: Re-inventing Rites of Passage* (Berkeley, CA: University of California Press).

Grinyer, A. (2002) 'The anonymity of research participants: assumptions, ethics and practicalities', *Social Research Update*, http://sru.soc.surrey.ac.uk/SRU36.html [accessed 01 March 2010].

Johannessen, H. and Lázár, I. (eds) (2006) *Multiple Medical Realities. Patients and Healers in Biomedical, Alternative and Traditional Medicine* (New York: Oxford).

Habenstein, R. (1954) *The American Funeral Director: A Study in the Sociology of Work*, Unpublished PhD thesis (University of Chicago).

Hale, R. (2005) 'The funeral of Angela Hale', *Natural Death Centre News and Views*, Autumn/Winter: 3.

Hallam, E. and Hockey, J. (2001) *Death, Memory and Material Culture* (Oxford: Berg).

Hallam, E., Hockey, J., and Howarth G. (1999) *Beyond the Body: Death and Social Identity* (London: Routledge).

Harris, C. (1987) 'The individual and society: a processual approach', in A. Bryman, B. Bytheway, P. Allatt, and T. Keil (eds) *Rethinking the Life Cycle* (London: Macmillan).

Hartig, K. and Dunn, K. (1998) 'Roadside memorials: interpreting new deathscapes in Newcastle, New South Wales,' *Australian Geographical Studies*, **36**: 5–20.

Hauser, M. (1987) 'Special aspects of grief after suicide', in E. J. Dunne, J. McIntosh, and K. Dunne-Maxim (eds) *Suicide and its Aftermath. Understanding and Counselling the Survivor* (London: W.W. Norton).

Hertz, R. ([1907] 1960) *Death and the Right Hand* (Aberdeen: Cohen and West).

Hetherington, K. (2004) 'Secondhandedness: consumption, disposal, and absent presence', *Environment and Planning D: Society and Space*, **22**: 157–173.

Hewitt, J. (1979) *Self and Society: A Symbolic Interactionist Social Psychology*, 2nd edn (Boston: Allyn and Bacon).

Ho, V. K. Y. (2000) 'Butchering fish and executing criminals', in G. Aijmer and J. Abbink (eds) *Meanings of Violence* (Oxford: Berg).

Hockey, J. (1990) *Experiences of Death, An Anthropological Account* (Edinburgh: Edinburgh University Press).

Hockey, J. (1996) 'The view from the West: reading the anthropology of non-western death ritual', in G. Howarth and P. C. Jupp (eds) *Contemporary Issues in the Sociology of Death, Dying and Disposal* (Basingstoke: Macmillan).

Hockey, J. (2001) 'Changing death rituals', in J. Hockey, J. Katz, and N. Small (eds) *Grief, Mourning and Death Ritual* (Buckingham: Open University Press).

Hockey, J. and Draper J. (2005) 'Beyond the womb and the tomb: identity, (dis)embodiment and the life course', *Body and Society*, **11** (2): 41–57.

Hockey, J. and James, A. (2003) *Social Identities across the Life Course* (Basingstoke: Macmillan).

Hockey, J., Kellaher, L., and Prendergast, D. (2007a) 'Of grief and well-being: competing conceptions of restorative ritualisation', *Anthropology and Medicine*, **14** (1): 1–14.

Hockey, J., Kellaher, L., and Prendergast, D. (2007b) 'Sustaining Kinship: ritualization and the disposal of human ashes in the United Kingdom', in M. Mitchell (ed.) *Remember Me: Constructing Immortality* (London: Routledge).

Hockey, J., Kellaher, L., and Prendergast, D. (2007c) 'La cremation et le devenir des cendres', *Ethnologie Francaise*, **37** (2): 295–304.

Hockey, J., Kelleher, L., and Prendergast, D. (2010) 'Ash disposal and the making of new memorial environments in the UK', in A. L. Molendijk, P. Post, and J. Kroesen (eds) *Holy Ground: Reinventing Ritual Space in Modern Western Culture* (Louvain, Belgium: Peeters Publishers).

Hockey, J., Penhale, B., and Sibley, D. (2001) 'Landscapes of loss: spaces of memory, times of bereavement', *Ageing and Society*, **21**: 739–757.

Hockey, J. Penhale, B. and Sibley, D. (2005) 'Environments of memory. Home space, later life and grief', in J. Davidson, L. Bondi, and M. Smith (eds) *Emotional Geographies* (Aldershot: Ashgate).

Holloway, M. (2007) *Negotiating Death in Contemporary Health and Social Care* (Bristol: Policy Press).

Horrox, R. (1999) 'Purgatory, prayer and plague: 1150–1380', in P. C. Jupp and C. Gittings (eds) *Death in England, An Illustrated History* (Manchester: Manchester University Press).

Hospice Friendly Hospitals Programme (2008) *Design and Dignity Guidelines for Physical Environments supporting End-of-Life Care* (Dublin: The Irish Hospice Foundation).

Howarth, G. (1996) *Last Rites: The Work of the Modern Funeral Director* (Amityville, NY: Baywood Publishing).

Howarth, G. (1997) 'Death on the road: the role of the English Coroner's Court in the social construction of an accident' in M. Mitchell (ed.) *The Aftermath of Road Accidents: Psychological, Social and Legal Consequences of an Everyday Trauma* (London: Routledge).

Howarth, G. (2000) 'Dismantling the boundaries between life and death', *Mortality*, **5** (2): 127–138.

Howarth, G. (2001a) 'Corpse', in G. Howarth and O. Leaman (eds) *Encyclopaedia of Death and Dying* (London: Routledge).

Howarth, G. (2001b) 'Grieving in Public', in J. L. Hockey, J. Katz, and N. Small (eds) *Grief, Mourning, and Death Ritual* (Buckingham: Open University).

Howarth, G. (2007) *Death and Dying: A Sociological Introduction* (Cambridge: Polity Press).

Hughes, B. (2000) 'Medicalised bodies', in P. Hancock, B. Hughes, E. Jagger, K. Paterson, R. Russell, E. Tulle-Winton, and M. Tyler (eds) *The Body, Culture and Society* (Buckingham: Open University Press).

Hussein, I. (2006) 'The cemetery as a garden: why a new understanding of the word 'cemetery' is necessary', *The Journal of the Institute of Cemetery and Crematorium Management* (Inc.), Autumn, London: ICCM: 13–16.

Hussein, I. and Rugg, J. (2003) 'Managing London's dead: a case of strategic policy failure', *Mortality*, **8** (2): 209–221.

Hutchins, E. (1994) *Cognition in the Wild* (Cambridge: MIT Press).

Illich, I. (1976) *Limits to Medicine* (London: Marion Boyars).

Ingold, T. (2000) *The Perception of the Environment, Essays on Livelihood, Dwelling and Skill* (London: Routledge).

Ingold, T. (2007) *Lines. A Brief History* (Abingdon: Routledge).

http://www.naturaldeath.org.uk/ [accessed 11 August 2008].

Jalland, P. (1999) 'Victorian death and its decline: 1850–1918', in P. C. Jupp and C. Gittings (eds) *Death in England, An Illustrated History* (Manchester: Manchester University Press).

James, A. and Hockey, J. (2007) *Embodying Health Identities* (Basingstoke: Palgrave).

James, P. and Noakes, T. (1994) *Hospital Architecture* (Harlow: Longman).

Janoff-Bulman, R. (1989) 'Assumptive worlds and the stress of traumatic events: applications of the schema', *Social Cognition*, **7** (2): 113–136.

Janoff-Bulman, R. (1992) *Shattered Assumptions: Towards a New Psychology of Trauma* (New York: The Free Press).

Jenkins, R. (2004) *Social Identity*, 2nd edn (London: Routledge).

Johnston, H. I. K. (2004) 'The natural way burial movement', *Connections: Scotland's Voice of Alternative Health Magazine*, **45**.

Jones, G. (2005) 'Mountain memorials', *The Scottish Mountaineer*, 28 (September) reprinted online: http://www.mountaineering-scotland.org.uk/nl/65b.html [accessed 11 August 2008].

Jones, L. (2000) *The Hermeneutics of Sacred Architecture* (Cambridge, Mass: Harvard University Press).

Jones, O. (2005) 'An ecology of emotion. Memory, self and landscape', in J. Davidson, L. Bondi, and M. Smith (eds) *Emotional Geographies* (Aldershot: Ashgate).

Jupp, P. C. and Walter, T. (1999) 'The healthy society: 1918–1998' in P. C. Jupp and C. Gittings (eds) *Death in England, An Illustrated History* (Manchester: Manchester University Press).

Justice, C. (1997) *Dying the Good Death, the Pigrimage to Die in India's Holy City* (Albany, New York: State University of New York Press).

Kan, S. (1989) *Symbolic Immortality, The Tlingit Potlatch of the Nineteenth Century* (Washington and London: Smithsonian Institution Press).

Katz, J. (2001) 'Introduction', in J. Hockey, J. Katz, and N. Small (eds) *Grief, Mourning and Death Ritual* (Buckingham: Open University Press).

Kauffman, J. (ed.) (2002) 'Safety and the assumptive world', *Loss of the Assumptive World. A Theory of Traumatic Loss* (London: Brunner Routledge).

Kellaher, L., Prendergast, D., and Hockey, J. (2005) 'In the shadow of the traditional grave', *Mortality*, **10** (4): 237–250.

Kellehear, A. (1984) 'Are we a "death denying" society? A sociological view', *Social Science and Medicine*, **18** (9): 713–723.

Kellehear, A. (2000) *Eternity and Me: The Everlasting Things in Life and Death* (Melbourne: Hill of Content).

Kellehear, A. (2007) *A Social History of Dying* (Cambridge and New York: Cambridge University Press).

King, A. (1998) *Memorials of the Great War in Britain: The Symbolism and Politics of Remembrance* (Oxford: Berg).

Klass, D., Silverman, P. R., and Nickman, S. L. (eds) (1996) *Continuing Bonds, New Understandings of Grief* (London: Taylor and Francis).

Komaromy, C. (2000) 'The sight and sound of death: the management of dead bodies in residential and nursing homes for older people', *Mortality*, **5** (3): 299–315.

Komaromy, C. (2002) 'The performance of the hour of death', in D. Clark and J. Hockley (eds) *Palliative Care for Older People in Care Homes* (Buckingham: Open University Press).

Komaromy, C. (2009) 'The State of Dying', in S. Earle, C. Komaromy and C. Bartholomew (eds.) *Death and Dying: A reader* (London: Sage/The Open University).

Kopytoff, I. (1986) 'The cultural biography of things: commoditization as process', in A. Appadurai (ed.) *The Social Life of Things: Commodities in Cultural Perspective* (Cambridge and New York: University Press).

Kouwenhaven, W. B. and Knickerbocker, G. (1960) 'Closed-chest cardiac massage', *Journal of the American Medical Association*, **173** (10): 1064–1067.

Kretschmer, A. (2000) 'Mortuary rites for inanimate objects: the case of *Hari Kuy*', *Japanese Journal of Religious Studies*, **27** (3–4): 379–404.

Kristeva, J. (1982) *Powers of Horror: An Essay on Abjection* (New York: Columbia University Press).

Kübler-Ross, E. (1970) *On Death and Dying* (London: Tavistock).

Laing and Buisson (2006) *Care of Elderly People, Market Survey* (2002–2009) (London: Laing and Buisson Publications).

Lakoff, G. and Johnson, M. (1980) *Metaphors We Live By* (Chicago: University of Chicago Press).

Lambert, D. (2006) *The Cemetery in a Garden: 150 years of the City of London Cemetery and Crematorium* (London: City of London).

Latour, B. ([1983] 1993) *We Have Never Been Modern* (New York: Harvester Wheatsheaf).

Law, J. (2004) *After Method: Mess in Social Science Research* (London: Routledge).

Lawler, J. (1991) *Behind the Screens. Nursing, Somology and the Problem of the Body* (Melbourne: Churchill Livingstone).

Lawton, J. (1998) 'Contemporary hospice care: the sequestration of the unbounded body and "dirty" dying', *Sociology of Health and Illness*, **20** (2): 121–143.

Lawton, J. (2000) *The Dying Process. Patients' Experiences of Palliative Care* (London: Routledge).

Layne, L. (2000) 'He was a real baby with baby things. A material culture analysis of personhood, parenthood and pregnancy loss', *Journal of Material Culture*, **5** (3): 321–345.

Layne, L. (2003) *Motherhood Lost: A Feminist Account of Pregnancy Loss in America* (London and New York: Routledge).

Lázár, I. and Johannessen, H. (eds) (2006) 'Epilogue – multiple medical realities: reflections from medical anthropology', *Multiple Medical Realities: Patients and*

Healers in Biomedical, Alternative and Traditional Medicine (Oxford: Berghahn Books).
Leder, A. (1990) *The Absent Body* (Chicago and London: University of Chicago Press).
Lefebvre, H. (1991) *The Production of Space* (Oxford: Blackwell).
Leonard, J. (1997) *How Conflicts Are Commemorated in Northern Ireland* (Belfast: Central Community Relations Unit).
Lewis, I. M. (1971) *Ecstatic Religion* (London: Pelican).
Lienhardt, G. (1961) *Divinity and Experience: The Religion of the Dinka* (Oxford: Clarendon Press).
Littlewood, J. (2001) 'Just an old fashioned love song or a "harlequin romance"?' in J. Hockey, J. Katz, and N. Small (eds) *Grief, Mourning and Death Ritual* (Buckinghamshire: Open University Press).
Lock, M. (1996) 'Death in technological time: locating the end of meaningful life', *Medical Anthropology Quarterly*, **10** (4): 575–600.
Lock, M. (2002) *Twice Dead. Organ Transplants and the Reinvention of Death* (Berkeley: University of California Press).
Loudon, J. C. ([1843] 1981) *On the Laying Out, Planting, and Managing of Cemeteries and on the Improvement of Churchyards. New edition with Introduction by James Stevens Curl* (Redhill, Surrey: Ivelet).
Lovell, A. (1997) 'Death at the beginning of life', in D. Field, J. Hockey, and N. Small (eds) *Death, Gender and Ethnicity*, (London: Routledge).
Lupton, D. (1994) *Medicine as Culture: Illness, Disease and the Body in Western Societies* (London: Sage).
Lupton, D. (2004) *Medicine as Culture*, 2nd edn (London: Sage).
MacConville, U. and McQuillan, R. (2005) 'Continuing the tradition: roadside memorials in Ireland', *Archaeology Ireland*, **19** (1): 26–30.
Macnaghten, P. and Urry, J. (1998) *Contested Natures* (London: Sage).
Marshall, P. (2002) *Beliefs and the Dead in Reformation England* (Oxford: Oxford University Press).
Marwit, S. J. and Klass, D. (1996) 'Grief and the role of the inner representation of the deceased', in D. Klass, P. R. Silverman, and S. L. Nickman (eds) *Continuing Bonds* (London: Taylor and Francis).
Maschio, T. (1994) *To Remember the Faces of the Dead* (Madison, WI: University of Wisconsin Press).
Mauss, M. ([1934] 1973) 'Techniques of the body', *Economy and Society*, **2**: 70–88.
Mauss, M. (1987) 'A category of the human mind: the notion of person; the notion of self', in M. Carrithers, S. Collins, and S. Lukes (eds) *The Category of the Person, Anthropology, Philosophy, History* (Cambridge: Cambridge University Press).
Mayor, S. (2007) 'Care of dying patients and safety dominate report on NHS complaints', *British Medical Journal*, **334**: 278.
McKay, S. (2006) 'North's divisions follow victims to their graves' (*The Irish Times*, September 14).
McManners, J. (1981) *Death and the Enlightenment: Changing Attitudes to Death Among Christians and Unbelievers in Eighteenth-Century France* (Oxford: Clarendon).
McNamara, B. (2001) *Fragile Lives: Death, Dying and Care* (Buckingham: Open University Press).

Mead, G. (1934) *Mind, Self and Society* (Chicago: University of Chicago Press).
Mellor, H. and Parsons, B. (2008) *London Cemeteries: An Illustrated Guide and Gazetteer*, 4th edn (Stroud: Sutton Publishing Limited).
Mellor, P. (1993) 'Death in high modernity: the contemporary presence and absence of death', in D. Clark (ed.) *The Sociology of Death* (Oxford: Blackwell).
Mellor, P. and Shilling, C. (1993) 'Modernity, self-identity and the sequestration of death', *Sociology*, **27** (3): 411–432.
Mental Health Act (2007) http://www.opsi.gov.uk/si/si-2008-index [accessed 3 December 2009].
Meyer, M. and Woodthorpe, K. (2008) 'The material presence of absence: a dialogue between museums and cemeteries', *Sociological Research Online*, **13** (5), http://www.socresonline.org.uk/13/5/1.html
Miles, M. and Huberman, A. (1994) *Qualitative Data Analysis: An Expanded Sourcebook*, 2nd edn (Thousand Oaks: Sage Publications).
Milton, K. (2002) *Loving Nature: Towards an Ecology of Emotion* (London: Routledge).
Ministry of Justice (2009) *Statistics on Deaths Reported to Coroners in England and Wales, 2008, Ministry of Justice Statistics Bulletin* (London: Ministry of Justice).
Mitchell, A. M., Gale, D. D., Garand, L., and Wesner, S. (2003) 'The use of narrative data to inform the psychotherapeutic group process with suicide survivors', *Issues in Mental Health Nursing*, 24: 91–106.
Mitford, J. (1963) *The American Way of Death* (New York: Simon and Schuster).
Mohammed, S. and Peter, E. (2009) 'Rituals, death and the moral practice of medical futility', *Nursing Ethics*, **16** (3): 292–302.
Mol, A. (2002) *The Body Multiple: Ontology in Medical Practice* (London: Duke University Press).
Morris, T. (2009) 'Unauthorized memorabilia'. Presentation and debate at the Institute of Cemetery and Crematorium Management Corporate Seminar (Cranfield University, May).
Mountaineering Council of Scotland (2008) *Mountain Memorials*, http://www.mountaineering-scotland.org.uk/access/memorials.html [accessed 22 January 2010].
Moyaert, P. (2007) *Iconen en beeldverering: godsdienst als symbolische praktijk* (Nijmegen: SUN).
Mozley, C., Sutcliffe, C., Eagley, H., Cordingley, L., Challis, D., Huxley, P., and Burns, A. (2004) *Towards Quality Care: Outcomes for Older People in Care Homes* (Hampshire: Ashgate).
Mulkay, M. (1993) 'Social death in Britain', in D. Clark (ed.) *The Sociology of Death*: (Oxford: Blackwell).
National statistics (2007) 'Death registrations', www.statistics.gov.uk/cci/nugget.asp?ID=949 [accessed 14 December 2009].
Naylor, M. (1989) *Funeral Rituals in a Northern City*, Unpublished PhD thesis (University of Leeds).
Needham, R. (1981) *Circumstantial Deliveries* (Berkeley: University of California Press).
Neimeyer, R. A. (1999). 'Narrative strategies in grief therapy', *Journal of Constructivist Psychology*, **12**: 65–85.
Neimeyer, R. A. (2000) 'Searching for the meaning of meaning: grief therapy and the process of reconstruction', *Death Studies*, **24**: 541–558.

Neuberger, J. (2004) *Dying Well: A Guide to a Good Death*, 2nd edn (Oxon: Radcliffe Publishing Ltd).

Neuberger, J. (2008) *Not Dead Yet: A Manifesto for Old Age* (London: Harper Collins).

Nic Neill, M. (1948) 'Wayside death cairns in Ireland', *Béaloideas, The Journal of the Folklore of Ireland Society*, **IML XVI**: 49–63.

Nisbet, R. A. (1969) *Social Change and History: Aspects of the Western Theory of Development* (New York: Oxford University Press).

O'Hear, A. (1998) 'Diana, queen of hearts: Sentimentality personified and canonised', in D. Anderson and P. Ullen (eds) *Faking it: The Sentimentalisation of Modern Society* (London: Social Affairs Unit).

O'Toole, P. and Were, P. (2008) 'Observing places: using space and material culture in qualitative research', *Qualitative Research*, **8** (5): 616–634.

Okely, J. (1983) *The Traveller Gypsies* (Cambridge: Cambridge University Press).

Otto, R. ([1917] 1924) *The Idea of the Holy* (Oxford: Oxford University Press).

Parkes, C. M. (1996) *Bereavement: Studies of Grief in Adult Life*, 3rd edn (London: Routledge).

Parkin, A. J. (1993) *Memory, Phenomena, Experiment and Theory* (Oxford: Blackwell).

Payne, M. (2006) *Narrative Therapy*, 2nd edn (London: Sage Publications).

Petersson, A. (2004) *The Presence of the Absent: Memorials and Places of Ritual* (Lund Institute of Technology: Dept of Architecture, unpublished dissertation).

Petersson, A. (2005) 'The Production of a Proper Place of Death', paper presented to the 7th International Conference on the Social Context of Death, Dying and Disposal, Bath.

Pfeffer, C. R., Martins, P., Mann, J., Sunkenberg, R. N., Ice, A., DaMore, J. P., Gallo, C., Kaprenos, I., and Jiang, H. (1997) 'Child survivors of suicide: psychosocial characteristics', *Journal of the American Academy of Child and Adolescent Psychiatry*, **36**: 65–74.

Pilcher, J. (1995) *Age and Generation in Modern Britain* (Oxford: Oxford University Press).

Pine, V. (1975) *Caretaker of the Dead: The American Funeral Director* (New York: Irvington Publishers Inc.).

Porter, R. (1999) *The Greatest Benefit to Mankind: A Medical History of Humanity from Antiquity to the Present* (London: Fontana Press).

Potts, T. J. (2007) 'Crying the wrong tears: floral tributes and aesthetic judgement', http://mlpa.nottingham.ac.uk/archive/00000072/ [accessed 30 September 2009].

Prendergast, D. (2005) *From Elder to Ancestor: Old Age, Death and Inheritance in Modern Korean Society* (Folkestone: Global Oriental).

Prendergast, D., Hockey, J., and Kellaher, L. (eds) (2006) 'Blowing in the wind? Identity, materiality, and the destinations of human ashes', *Journal of the Royal Anthropological Institute*, **12** (4): 881–898.

Prior, L. (1989) *The Social Organisation of Death: Medical Discourse and Social Practices in Belfast* (Basingstoke: Macmillan).

Range, L. M. and Niss, N. M. (1990) 'Long-term bereavement from suicide, homicide, accidents and natural deaths', *Death Studies*, **14**: 423–433.

Raphael, B. (1984) *The Anatomy of Bereavement* (London: Routledge).

Rapport, N. and Overing, J. (2000) *Social and Cultural Anthropology: The Key Concepts* (London: Routledge).

Realdania Fund (2006) *Programme for the Good Hospice in Denmark*. English translation available, www.hospiceforum.dk [accessed 22 January 2010].

Relph, E. (1976) *Place and Placelessness* (London: Pion Ltd).

Richert, A. J. (1999) 'Some practical implications of integrating narrative and humanistic/existential approaches to psychotherapy', *Journal of Psychotherapy Integration*, **9** (3): 257–278.

Riches, G. and Dawson, P. (1998) 'Lost children, living memories: the role of photographs in processes of grief and adjustment among bereaved parents', *Death Studies*, **22**: 121–140.

Roach, J. (1996) *Cities of the Dead: Circum-Atlantic Performance* (New York: Columbia University Press).

Robben, A. (ed.) (2004) *Death, Mourning and Burial* (Oxford: Blackwell).

Romanucci-Ross, L., Moerman, D. E., and Tancredi, L. R. (eds) (1997) 'Preface', *The Anthropology of Medicine: From Culture to Method*, 3rd edn (Westport, CT: Bergin and Garvey).

Rosenblatt, P. (1996) 'Grief that does not end', in D. Klass, P. R. Silverman and S. L. Nickman (eds) *Continuing Bonds: New Understandings of Grief* (London: Taylor and Francis).

Rowling, L. (1999) 'Being in, being out, being with: affect and the role of the qualitative researcher in loss and grief research', *Mortality*, **4** (2): 167–181.

Rugg, J. (1999) 'From reason to regulation: 1760–1850', in P. C. Jupp and C. Gittings (eds) *Death in England, An Illustrated History* (Manchester: Manchester University Press).

Rugg, J. (2000) 'Defining the place of burial: what makes a cemetery a cemetery?' *Mortality*, **5** (3): 259–275.

Rugg, J. (2006) 'Lawn cemeteries: the emergence of a new landscape of death', *Urban History*, **33** (2): 213–233.

Russac, R. J., Steighner, N. S., and Canto, A. I. (2002) 'Grief work versus continuing bonds: a call for paradigm integration or replacement?' *Death Studies*, **26**: 463–478.

Rutherford, S. (2008) *The Victorian Cemetery* (Oxford: Shire Publications).

Saint, A. (1987) *Towards a Social Architecture: Role of School Buildings in Post-war England* (London: Yale University Press).

Sanders, C. M. (1999) *Grief: The Mourning After: Dealing with Adult Bereavement*, 2nd edn (New York: John Wiley and Sons Inc.).

Santino, J. (2001) *Signs of War and Peace* (London: Palgrave Macmillan).

Santino, J. (ed.) (2006) *Spontaneous Shrines and the Public Memorialization of Death* (Basingstoke: Palgrave Macmillan).

Sarkis, H. with Allard, P., and Hyde, T. (2001) *Le Corbusier's Venice Hospital and the Mat Building Revival* (Munich: Prestel).

Sather, C. (2003) 'Transformations of self and community in Saribas Iban' in W. Wilder (ed.) *Journeys of the Soul, Anthropological Studies of Death, Burial and Reburial Practices in Borneo* (Borneo: Borneo Research Council).

Saunders, C. (2006) *Selected Writings 1958–2004* (Oxford: Oxford University Press).

Schäfer, C. (2007) 'Post-mortem personalization: pastoral power and the New Zealand funeral director', *Mortality*, **12** (1): 4–21.

Schama, S. (1995) *Landscape and Memory* (London: HarperCollins).
Scheper-Hughes, N. and Lock, M. M. (1987) 'The mindful body: a prolegomenon to future work in medical anthropology', *Medical Anthropology Quarterly*, **1** (1): 6–41.
Schlereth, T. J. (ed.) (1982) 'Material culture studies in America. 1876–1976', *Material Culture Studies in America* (Nashville, TN: American Association for State and Local History).
Seale, C. (1998) *Constructing Death: The Sociology of Dying and Bereavement* (London: Sage).
Seremetakis, C. N. (1991) *The Last Word: Women, Death and Divination in Inner Mani* (Chicago: Chicago University Press).
Seremetakis, C. N. (1994) *The Senses Still: Perception and Memory as Material Culture in Modernity* (Chicago: University of Chicago Press).
Sethi, S. and Bhargava, S. C. (2003) 'Child and adolescent survivors of suicide', *Crisis*, **24** (1): 4–6.
Seymour, J. E. (1998) *Caring for Critically Ill People: A Study of Death and Dying in Intensive Care*, Unpublished PhD thesis (University of Sheffield).
Shepherd, D. and Barraclough, B. M. (1974) 'The aftermath of suicide', *British Journal of Psychiatry*, **2**: 600–603.
Shilling, C. (2003) *The Body and Social Theory*, 2nd edn (London: Sage Publications).
Shiner, M., Scourfield, J., Fincham, B., and Langer, S. (2009) 'When things fall apart: gender and suicide across the life-course', *Social Science and Medicine*, **69**: 738–746.
Sidaway, J. and Maddrell, A. (eds) (2010) *Deathscapes: Spaces for Death, Dying and Bereavement* (Farnham: Ashgate).
Sidell, M., Katz, J., and Komaromy, C. (1997) *Death and Dying in Residential and Nursing Homes for Older People: Examining the Case for Palliative Car*e (London: Department of Health).
Sidell, M. and Komaromy, C. (2003) 'Who dies in care homes for older people?' in J. Katz and S. Peace (eds) *End of Life in Care Homes: A Palliative Care Approach* (Oxford: Oxford University Press).
Silverman, E., Range, L., and Overholser, J. (1994–95) 'Bereavement from suicide as compared to other forms of bereavement', *Omega*, **30** (1): 41–51.
Sloane, D. (1991) *The Last Great Necessity: Cemeteries in American history* (Baltimore: Johns Hopkins University Press).
Smale, B. (1985) *Deathwork: A Sociological Analysis of Funeral Directing*, Unpublished PhD thesis (University of Surrey).
Small, N. (2001) 'Theories of grief: a critical overview', in J. Hockey, J. Katz, and N. Small (eds) *Grief, Mourning and Death Ritual* (Buckingham: Open University Press).
Smith, R. J. (1974) *Ancestor Worship in Contemporary Japan* (Stanford: Stanford University Press).
Sokol, D. K. (2009) 'The death of DNR', *British Medical Journal*, **338**: 1723.
Speyer, J. (2001) 'Nicholas Albery's Woodland Funeral and Memorial Celebration', in N. Albery, S. Wienrich, N. Temple, and R. Bowen (eds) *Progressive Endings: Changing Attitudes to Death and Dying* (London: Natural Death Centre).
Speyer, J. and Wienrich, S. (2003) *The Natural Death Handbook*, 4th edn (London: Rider).

Stirling, F. M. (2009) *Grave re-use: a feasibility study*, Unpublished PhD thesis (University of Sheffield).

Strathern, M. (1988) *The Gender of the Gift: Problems with Women and Problems with Society in Melanesia* (Berkeley: University of California Press).

Stroebe, M. S., Hanson, R. O., Stroebe, W. and H. Schut (eds) (2001) *Handbook of Bereavement Research and Practice: 21st Century Perspectives* (Washington, DC: American Psychological Association).

Stroebe, M. and Schut, H. (2005) 'To continue or relinquish bonds; a review of consequences for the bereaved', *Death Studies*, **29**: 477–494.

Stroebe, M., Gergen, M., Gergen, K., and Stroebe, W. (1996) 'Broken hearts or broken bonds?' in D. Klass, P. R. Silverman, and S. L. Nickman (eds) (1996) *Continuing Bonds* (London: Taylor and Francis).

Sudnow, D. (1967) *Passing On: The Social Organisation of Dying* (Englewood Cliffs, NJ: Prentice-Hall).

Synnott, A. (1992) 'Tomb, temple, machine and self: the social construction of the body', *The British Journal of Sociology*, **43** (1): 79–110.

Synott, A. (1993) *The Body Social: Symbolism, Self and Society* (London: Routledge).

Tarlow, S. (2000) 'Landscapes of memory: the nineteenth-century garden cemetery', *European Journal of Archaeology*, **3** (2) 217–239.

Tay, C. (2009) 'Drivers' perceptions and reactions to roadside memorials', *Accident Analysis and Prevention*, **41**: 663–669.

Taylor, D. (2003) *The Archive and the Repertoire: Performing Cultural Memory in the Americas* (Durham, NC: Duke University Press).

Telegraph (2001) 'Dearly departed – but never far away' (*Daily Telegraph*, December 8), http://www.telegraph.co.uk/education/main.jhtml?xml=/education/2001/12/14/tepgrave17.xml [accessed 13 August 2008].

The Home Office (2004) *Burial Law and Policy in the 21st Century: The Need for a Sensitive and Sustainable Approach* (London: The Home Office).

The Stationery Office (2001) *Environment, Transport and Regional Affairs Committee, Cemeteries, 8th Report, Vol. 1, Report and Proceedings of The Committee* (London: the Stationery Office Limited).

The Stationery Office (2001) *Environment, Transport and Regional Affairs Committee, Cemeteries,* **I and II** (London: House of Commons).

Thomas, K. (1983) *Man and the Natural World: Changing Attitudes in England 1500–1800* (London: Allen Lane).

Thompson, J. W. (2002) 'A natural death', *Landscape Architecture*, **92** (10): 74–79.

Tilley, C. (2002) 'Metaphor; materiality and interpretation', in V. Buchli (ed.) *The Material Culture Reader* (Oxford: Berg).

Timmermans, S. (1999) *Sudden Death and the Myth of CPR* (Philadelphia: Temple University Press).

Timmermans, S. (2002) 'Hearts too good to die: Claude S Beck's contributions to life-saving', *Journal of Historical Sociology*, **14** (1): 108–131.

Timmermans, S. (2006) *Postmortem: How Medical Examiners Explain Suspicious Deaths* (Chicago and London: University of Chicago Press).

Townsend, P. (1962) *The Last Refuge: A Survey of Residential Institutions and Homes for the aged in England and Wales* (London: Routledge and Kegan Paul).

Tulle-Winton, E. (2000) 'Old bodies', in P. Hancock, B. Hughes, E. Jagger, K. Paterson, R. Russell, E. Tulle-Winton, and M. Tyler (eds) *The Body, Culture and Society* (Buckingham: Open University Press).

Turner, B. S. (1992) *Regulating Bodies. Essays in Medical Sociology* (London: Routledge).
Turner, B. S. (1995) *Medical Power and Social Knowledge*, 2nd edn (London: Sage).
Turner, V. (1969) *The Ritual Process* (London: Penguin Books).
Turner, V. (1975) *Dramas, Fields, and Metaphors: Symbolic Action in Human Society* (New York: Cornell University Press).
Tylor, E. B. (1958) *Primitive Culture*, Vol. 1 (New York: Harper).
Urbina, I. (2006) 'As roadside memorials multiply, a second look' (*New York Times*, February 6).
Valentine, C. (2007) *Bereavement and Identity: Making Sense of Bereavement in Contemporary British Society*, Unpublished PhD thesis (University of Bath).
Valentine, C. (2008) *Bereavement Narratives. Continuing Bonds in the Twenty-First Century* (London: Routledge).
Van Delden J. J. M. (2000) 'Do-not-resuscitate decisions', in D. Dickenson, M. Johnson, and J. S. Katz (eds) *Death, Dying and Bereavement* (London: Sage).
Van der Woning, M. (1997) 'Should relatives be invited to witness a resuscitation attempt. A review of the literature', *Accident and Emergency Nursing*, **5**: 215–218.
Van Gennep, A. ([1909] 1960) *The Rites of Passage* (Chicago: University of Chicago Press).
Van Reusel, W. (1999) *Leven of Dood: Over Moeilijke Grenzen* (Leuven: Davidsfonds Uitgeverij).
Vanderstraeten, R. (2009) 'Modes of individualisation at cemeteries', *Sociological Research Online*, **14** (4), http://www.socresonline.org.uk/14/4/10.html
Verderber, S. and Refuerzo, B. J. (2006) *Innovations in Hospice Architecture* (London: Taylor and Francis).
Verdery, K. (1999) *The Political Lives of Dead Bodies: Reburial and Postsocialist Change* (New York: Columbia University Press).
Vergunst, J. L. (2008) 'Taking a trip and taking care in everyday life', in T. Ingold and L. Vergunst (eds) *Ways of Walking. Ethnography and Practice on Foot* (Farnham: Ashgate).
Vincent, G. (1991) 'A history of secrets?' in A. Prost and G. Vincent (eds) *A History of Private Life. V. Riddles of Identity in Modern Times* (Cambridge, MA: The Belknap Press of Harvard University Press).
Wagner, R. (1991) 'The fractal person, in M. Godelier and M. Strathern (eds) *Big Men and Great Men: Personifications of Power in Melanesia* (Cambridge: Cambridge University Press).
Wagner-Pacifici, R. and Schwartz, B. (1991) 'The Vietnam veterans memorial: commemorating a difficult past', *American Journal of Sociology*, **97**: 376–420.
Walter, T. (1990) *Funerals and How to Improve Them* (London: Hodder and Stoughton).
Walter, T. (1991) 'Modern death: taboo or not taboo?' *Sociology*, **25** (2): 293–310.
Walter, T. (1994) *The Revival of Death* (London: Routledge).
Walter, T. (1996) 'A new model of grief: bereavement and biography', *Mortality*, **1** (1): 7–25.
Walter, T. (1997) 'Emotional reserve and the English way of grief', in K. Charmaz, G. Howarth, and A. Kellehear (eds) *The Unknown Country: Death in Australia, Britain, and the USA* (Basingstoke: Palgrave Macmillan).
Walter, T. (1999) *On Bereavement. The Culture of Grief* (Buckingham: Open University Press).

Walter, T. (2005) 'Three ways to arrange a funeral: mortuary variation in the modern West', *Mortality*, **10** (3): 173–92.

Walter, T. (2008) 'The new public mourning', in M. Stroebe, R. Hansson, W. Stroebe, and H. Schut (eds) *Handbook of Bereavement Research and Practice: 21st Century Perspectives* (Washington, DC: American Psychological Association).

War Graves of the Empire (1928) Reprinted from the Special Number of The Times for 10 November 1928 (London: Times Publishing Company).

Ward, D. (2006) *Hospice and Palliative Care Directory, United Kingdom and Ireland 2006* (London: Help the Hospices).

Warner, W. L. (1959) *The Living and the Dead* (Chicago: University of Chicago Press).

Wertheimer, A. (2001) *A Special Scar. The Experiences of People Bereaved by Suicide*, 2nd edn (Hove: Brunner-Routledge).

West, K. (2006) 'Memorial choice', *Memorial Awareness Board* (London: Houses of Parliament unpublished paper).

West, P. (2004) *Conspicuous Compassion: When Sometimes It Really is Cruel to be Kind* (London: Civitas).

Williams, R. (1990) *A Protestant Legacy: Attitudes to Death and Illness Among Older Aberdonians* (Oxford: Clarendon Press).

Willmott, H. (2000) 'Death. So what? Sociology, sequestration and anticipation', *The Sociological Review* **48** (4): 649–665.

Winter, J. (1995) *Sites of Memory, Sites of Mourning. The Great War in European Cultural History* (Cambridge: Cambridge University Press).

Witoszek, N. and Sheeran, P. (1994) *The Irish Funerary Tradition*. Research Report 4 (University College Galway: Social Science Research Centre).

Wojtkowiak, J. and Venbrux, E. (2007) 'The living and the dead: house shrines in the Netherlands', Paper given to the 8th International Conference on the Social Context of Death, Dying and Disposal, Bath.

Woodthorpe, K. (2007) *Negotiating ambiguity and uncertainty in the contemporary cemetery landscape*, Unpublished PhD thesis (University of Sheffield).

Woodthorpe, K. (2010) 'Buried bodies in an East London Cemetery: re-visiting taboo', in J. Sidaway and A. Maddrell (eds) *Deathscapes: Spaces for Death, Dying and Bereavement* (Farnham: Ashgate).

Worden, W. (1991) *Grief Counselling and Grief Therapy*, 2nd edn (London: Routledge).

Worpole, K. (1997) *The Cemetery in the City* (Stroud: Comedia).

Worpole, K. (2000) *Here Comes the Sun: Architecture and Public Space in 20th Century European Culture* (London: Reaktion Books).

Worpole, K. (2003) *Last Landscapes: The Architecture of the Cemetery in the West* (London: Reaktion Books).

Worpole, K. (2008) 'Hospice friendly hospitals programme', *Design and Dignity Guidelines for Physical Environments supporting End-of-Life Care* (Dublin: The Irish Hospice Foundation).

Worpole, K. (2009) *Modern Hospice Design: The Architecture of Palliative Care* (London: Routledge).

Young, Katherine (2002) 'The memory of the flesh', *Body and Society*, **8** (3): 25–48.

Zeldin, Theodore (1995) *An Intimate History of Humanity* (London: Vintage).

Index